Gratifying Transitions

Key Issues in Climate Change and Sustainability: Ethics, Politics and Policy

Series Editor

Idil Boran, Xavier Landes, & Morten Ebbe Juul Nielsen

Mission Statement

The book series *Key Issues of Climate Change: Ethics, Politics and Policy* provides quality, research-based monographies that address specific issues raised by climate change (e.g., immigration, geoengineering, corporate social responsibility). The series will offer books that provide an encompassing view of a given issue with a content that is accessible to a broad audience (students, lecturers, researchers, policymakers, citizens). Ultimately, the series will provide the knowledge and conceptual tools for nurturing debates and the search for solutions in terms of climate change mitigation and adaptation, as well as for issues that emerge as a consequence (e.g., migration). In short, the book series will create a forum for discussing concrete steps and potential consequences of living in a more uncertain world due to climate change.

Advisory Board Members

Recent Titles in the Series

Gratifying Transitions: Maintaining Wellbeing in the Face of Climate Change, by Søren Harnow Klausen

Engineering and Governing the Climate: Ethical and Political Issues, by Xavier Landes

Gratifying Transitions

Maintaining Wellbeing in the Face of Climate Change

Søren Harnow Klausen

BLOOMSBURY ACADEMIC
NEW YORK • LONDON • OXFORD • NEW DELHI • SYDNEY

BLOOMSBURY ACADEMIC
Bloomsbury Publishing Inc, 1359 Broadway, 12th floor, New York, NY 10018, USA
Bloomsbury Publishing Plc, 50 Bedford Square, London, WC1B 3DP, UK
Bloomsbury Publishing Ireland, 29 Earlsfort Terrace, Dublin 2, D02 AY28, Ireland

First published in the United States of America 2026

Library of Congress Cataloging-in-Publication Data Available

ISBN:	HB:	978-1-6669-1441-2
	ePDF:	979-8-21626-679-2
	eBook:	978-1-9787-6414-9

Series: Key Issues in Climate Change and Sustainability: Ethics, Politics and Policy

Typeset by Integra Software Services Pvt. Ltd.

For product safety related questions contact productsafety@bloomsbury.com.

Contents

1 Climate Change, Life Changes, and Wellbeing 1
- 1.1 The Role of Wellbeing in Responding to Climate Change 4
- 1.2 Wellbeing and Change 6
- 1.3 Wellbeing, Motivation, and Action 7
- 1.4 Views on the Relationship between Climate Change, Climate Action, and Wellbeing 8
- 1.5 Why Not Gamble on Green Growth? 11
- 1.6 Why Not Just Politics? 13
- 1.7 Wellbeing for Whom? 15
- 1.8 An Interdisciplinary, Philosophy-Driven Approach 16

2 Life in the Future 21
- 2.1 The Ubiquity of Change 21
- 2.2 Change for Better or Worse? 23
- 2.3 Going through Changes 25
- 2.4 Quantity and Quality of Change 27
- 2.5 Living with Climate Change 35
- 2.6 Living with Mitigation 39

3 A Theory of Wellbeing for Green Transition 49
- 3.1 A Plethora of Theories and a Provisional Assessment 50
- 3.2 Problems with Subjective Wellbeing? Immeasurability, Growth Ideology, and Adaptation 55
- 3.3 Virtues of Subjective Wellbeing 59
- 3.4 Growth Ideology and Individualism? 61
- 3.5 Why Not a Needs Approach? 63
- 3.6 Genuinely Basic Needs? 72
- 3.7 Adaptation for Good and for Bad 77
- 3.8 Richer and Poorer Lives 83
- 3.9 Getting the Most Out of It: Cultivation as the Key? 86
- 3.10 Putting It Together: A Theory of Wellbeing for Green Transition 89

4 Living with Change 97
4.1 Remaining, Losing, and Becoming Yourself 98
4.2 The Shape, Meaning, and Management of Life 104
4.3 Temporal Styles and Agents of Change 108
4.4 Keeping the Good Things Alive—Variation as the Key to Wellbeing? 110
4.5 Values and Impacts of Personal Change 115

5 Shaping Behavior: Self-Control, Cultivation, and Emulation 127
5.1 The Bad News: How Not to Change Behavior 127
5.2 The (Somewhat) Better News: How We *Might* Change Behavior 129
5.3 Limits of Behavioral Design 134
5.4 *Bildung* to the Rescue? 137
5.5 Getting It Right: *Bildung* and the Situationist Challenge 143
5.6 *Bildung* and Behavior Change as Social Process 149
5.7 Return of the Environmental Humanities? 153

6 Acceptable Transitions 157
6.1 Only Soft Interventions? 158
6.2 Goodbye to Liberalism? 160
6.3 Harm and Sustainability 164
6.4 Learning from the Pandemic 166
6.5 Realism, but Not Too Much: Balancing Feasibility Concerns and the Role of Hope 169
6.6 Moral Aesthetics 175
6.7 Unjustified Interventions? 178

7 Changemaking: Gratifying Transitions and Strategies for Green Transition 187
Coda: Living with Change in an Uncertain World 199

References 201
Index of Names 227
Index of Subjects 228
About the Author 229

1

Climate Change, Life Changes, and Wellbeing

You Must Change Your Life.

Rainer Maria Rilke*

Mounting an adequate response to climate change is arguably the most important task facing human civilization in the twenty-first century. It will very likely require people of the Global North, and probably people all around the globe, to change their way of life substantially. This book explores how and how far such changes are possible and what they will mean to human wellbeing. Because wellbeing is a contested notion, and existing theories of wellbeing are not well suited to handle the case of climate change action, a significant part of the book is about how to develop a more appropriate understanding of wellbeing. Although this amounts to a domain-specific theory of wellbeing, considering the case of climate change holds lessons for wellbeing theory and its application more broadly. It can teach us something about the possibility and consequences of life changes in general, and how to think about wellbeing during and after such changes.

The book is not about climate ethics as such. It is not directly concerned with questions of responsibility, obligations, equality, or international or intergenerational justice, and it remains mostly neutral to different ethical and political theories and principles. It is concerned with the *prudential* aspect of climate change and climate change action: whether and how we can live a good and satisfying life under constraints caused by climate change (including self-imposed constraints aimed at mitigating and adapting to climate change). In doing so, however, it deals with a central, but insufficiently acknowledged, precondition of climate change ethics. For questions of wellbeing are central to climate ethics in at least two ways. The expected impact on wellbeing is crucial

* *Archaic Torso of Apollo*, in *Ahead of All Parting: Selected Poetry and Prose of Rainer Maria Rilke*, transl. Stephen Mitchell. Modern Library 1995.

to the assessment and choice of actions and policies. And subjective, that is, felt, perceived, estimated, or expected, wellbeing is crucial to their feasibility. Hence wellbeing has both normative and instrumental significance to the question of how to respond adequately to climate change.

The book is motivated by a recognition that strategies for pro-environmental behavioral change need to be both imaginative and realistic. Much philosophical value theory tends to be conservative and unimaginative, underestimating the power of human creativity and the diversity of ways in which it is possible to live a satisfactory life. People are generally able to change and adapt to new circumstances, more than they realize. At the same time, there are serious psychological, social, and infrastructural obstacles to effectuating lifestyle changes. Many recommendations for simpler or more responsible ways of living have failed to take these obstacles into account.

The current situation is frustrating not because there is a lack of knowledge of climate change and its consequences, nor because of any shortage of good will or pro-environmental beliefs and attitudes. At the time of writing, many governments throughout the world have set climate change mitigation very high on the agenda. Nevertheless, it is widely agreed that the policies are inadequate and their implementation far too slow. There are even genuine backlashes, for example, when it comes to wind energy expansion, which in some of the countries with the strongest support for green policies (like Germany and Denmark) has recently stalled. This illustrates the need for a realistic approach, which takes seriously the psychological and sociological barriers and limitations to climate action and concentrates on actions and measures that are likely to make a difference in the fairly short term.

This realism signals a departure from the "environmental humanities," as these are usually understood (Cohen, 2021). I am not concerned with analyzing and altering our relationship with nature, with cultivating awe or respect or changing people's mindset. For not only will such a change in fundamental ways of thinking—assuming that it is possible, and desirable,[1] at all—take much longer time than we have, considering the urgency of the climate crisis. As we shall see, knowledge and attitudes seem to matter little to the kind of behavioral change that is needed to respond to climate change. We need action, not just thinking, and unfortunately, the link between thinking and action is not very firm.

Still, my approach is distinctively humanistic. It focuses on the human interface in climate change action: the human (often all too human) factors and tendencies that mediate between what we know and can (in principle, also in terms of our technological possibilities) and what we eventually end up doing. And I am not only concerned with "what works" in the sense that it can somehow bring about pro-environmental behavior, but also with whether it is acceptable. So ethics does play a role, albeit in the form *of climate change intervention ethics*, rather than climate change ethics more generally. It turns out that alleged mechanisms of behavior change are themselves sensitive to cultural contexts and values, so behavioral design alone will not do, either. I will argue that we need green transition with a human face, for both ethical and instrumental reasons. But I reckon that some will balk at the very idea of turning to behavioral science, finding it insufficiently humane or respectful to focus on *behavior* rather than on (autonomous, deliberate, ethically motivated, etc.) human *action*.

To this I am inclined to say with Chinese reformer Deng Xiaoping that it does not matter whether a cat is white or black, as long as it catches mice (Li, 1977). We need quick and efficient responses to climate change and cannot afford to be picky. That said, I have already made clear that I think behavior change must be brought about in an acceptable and sensitive way. I am not supporting manipulation or coercion. And I will argue that my approach is in line with the eminently humanistic ideal of self-cultivation or *Bildung*, which, rightly understood, is also concerned with behavior as much as it is concerned with action and promises to overcome the dichotomy between the two.

So although I am critical of popular notions of environmental *Bildung* or literacy, or of cultivating "green virtues" (Jamieson, 2014; Jordan & Kristjánsson, 2017), which seem to me overly intellectualist and idealistic, I share the belief that a process of cultivation, on both a collective and individual level, is needed for the good intentions to be realized before it is too late. And though I will maintain that changing people's overall mindset or establishing a closer relationship with nature is too difficult and inefficient when speedy changes are needed, at least as our main strategy, I will eventually concur that we should work at changing our personalities. I also see a point in fostering contact with and appreciation of nature, but mainly because of its aesthetic qualities and propensity for enhancing wellbeing. I even come to acknowledge a role for fostering knowledge and attitudes. But since this is mostly knowledge about behavior change, my view still differs markedly from the one that is typical of the environmental humanities and much of the natural science community, which also seems to think that getting the facts right will make people change their way of life.

1.1 The Role of Wellbeing in Responding to Climate Change

Knowledge about the consequences for human wellbeing of both climate change itself and mitigation and adaption is crucial to the choice of strategies and policies. We need it to justify specific measures and interventions. But we also need it to assess and improve their feasibility. Although human behavior is shaped by other factors than the concern for wellbeing, behavioral change can be quicker and smoother if it is perceived as sufficiently pleasant and rewarding. Otherwise promising plans for climate action are likely to falter if they do not take motivating and inhibiting factors, like felt and expected rewards and costs, into account.

Yet somewhat surprisingly, the role of wellbeing in climate change response has not been discussed extensively in neither the academic literature nor public discourse. Sometimes it is almost completely neglected. A standard work in climate ethics, like Gardiner's *A Perfect Moral Storm*, does not mention wellbeing at all and only marginally considers the potential impact on quality of life of climate change action (Gardiner, 2006). It also ignores the widely endorsed principle that "ought implies can": we can only require people to do what they are actually able to do.[2] Even if this may not hold without qualifications, it remains beyond discussion that practical limitations constrain moral obligations (Buckwalter, 2020).

In other cases, the role of wellbeing does seem to be duly acknowledged, but is then treated as obvious or trivial, rather than something which merits special concern. The urgency of climate change action is often explained by pointing to the disastrous effects of unmitigated climate change, with explicit or implicit reference to the living conditions of future generations (UN, 2023a). Moreover, it is commonly assumed either that the necessary changes will be very burdensome (Broom, 2022) or that we face something like a potential win-win situation, because the changes may actually turn out to be good for people, even if they do not themselves feel or expect them to be so (UN, 2023b). Both are serious hypotheses which I will consider at length. But the mere fact that both views are widely held, despite being contradictory, should be enough to show that none of them can be obviously correct. For they seem to rest on fundamentally different views on what constitutes wellbeing. Does it consist in satisfaction of current preferences, which may be impossible if we are to live sustainably? Or does it consist in meeting objective standards for a good life, which may require one to live less materialistically, anyhow, regardless of how one feels about it now?

This shows that philosophical and psychological questions about the nature of wellbeing are directly relevant to climate ethics and climate change policy. We cannot deliberate about the merits of different strategies without some common ground as to what can be said to make human lives go better or worse—or, at the very least, without making the views of wellbeing involved sufficiently explicit. We cannot even take completely for granted that unmitigated or half-heartedly mitigated climate change will make it impossible for future generations to achieve sufficient wellbeing. This is not some sophisticated form of climate skepticism, which accepts the reality of climate change, but denies the seriousness of its effects. It has been noted by philosophers who are deeply concerned about climate change but acknowledge the tendency of humans to adapt to new and seemingly adverse circumstances and so maintain a significant degree of subjective (that is, self-reported) wellbeing (Mulgan, 2015; O'Neill, 2017). Hence even a seemingly incontestable point like the harmfulness of climate change turns out to depend on assumptions about the nature and conditions of wellbeing.

The International Panel of Climate Change, IPCC, has devoted a whole chapter of its most recent report to the consequences of climate change for health and wellbeing. Yet it is much less conclusive than might be expected. It does amass evidence that climate change currently causes a wide range of harms and is likely to cause still more harms in the absence of adequate responses. But it also acknowledges that "climate resilient development has a strong potential to generate substantial co-benefits for health and well-being" (Cissé, 2022). Apart from being, understandably, impressionistic—the chapter is stitched together from very many different contributions from experts working with different approaches to wellbeing and health—it is also quite reserved in its judgments about the eventual consequences of various courses of action or inaction. I think this reflects more than the practical limitations of the IPCC's way of working and communicating. As I will argue later, it may be difficult to predict our future levels of wellbeing, even though many of the facts regarding both climate change and the necessary responses are clear—*and* worrisome—enough. Or to put it differently, one of the few things we can say with some certainty about the impact on wellbeing of climate change, and our response to it, is that it is very difficult to know in advance.

1.2 Wellbeing and Change

Responding to climate change is a task of such urgency and enormous importance that it should occupy most of our energy and attention, and also be a prime concern for philosophers (Brownstein et al., 2021). Still, it is not my only reason for taking up the topic of wellbeing and transitions. For this is a topic that has generally been neglected, not just in the case of climate change. Wellbeing theory, both in philosophy and psychology, has tended to be static. It has been mostly concerned with enumerating basic goods which people should try to possess (pleasure, absence of pain, autonomy, knowledge, love, friendship, etc.) or describing an ideal state of living (flourishing, thriving, living in accordance with one's nature or overall preferences, etc.). In contrast, it has been little concerned with the path to wellbeing—how to obtain the basic goods or achieve the ideal state—or with the consequences for wellbeing of trying to obtain the basic goods or getting into the ideal state (Klausen, 2018b).

This is a significant omission. Not only are prudential norms also subject to an "ought-implies-can" principle (of some sort)—the best for me must be the best I am capable to do or be. The quality of the transition process should also be included in the overall normative assessment. It might be said that this is already entailed by a theory like hedonism. Perhaps it is somehow entailed by all major theories of wellbeing and prudential rationality. Even so, it remains a striking fact that the theories either do not mention the process aspect at all or (like hedonism) sideline it as a merely "empirical" or "practical" question or recognize it as something that should "of course" be included in the overall hedonic calculus but is not a subject of theorizing in its own right.

A parallel limitation besets climate ethics. Jamieson paints a highly attractive, but also idealized picture of a "virtuous green" who possesses a number of virtues relevant to meeting climate change (Jamieson, 2014). It would indeed be good if such people would become plentiful, and it might solve the climate change problem entirely if they were to become dominant. But little of what Jamieson says in his more realistic assessment of biological, psychological, and institutional obstacles to climate action seems to provide any reason to believe that this is likely to happen. Nor does Jamieson offer any suggestions as to how to inculcate the virtues on which the future of mankind allegedly depends.

The need to make theories of wellbeing more sensitive to change is further highlighted by findings from empirical psychology. Reported subjective wellbeing is especially sensitive to changes in life condition (Besser, 2020; Railton, 2021). Some have gone so far as to claim that change is all that there is to achieving wellbeing—that in the absence of perceptible changes, the presence of allegedly supportive factors matters little to how well people feel they are doing. This is no doubt an exaggeration, but it remains a fact that without some elements of change or at least variation in what they experience, people have difficulties maintaining high subjective wellbeing (Sheldon et al., 2012), just as life with perceived changes to the worse is experienced as particularly bad (Lucas, 2007).

Moreover, changes and transitions are a fundamental and centrally important part of human lives. This has been widely, but only superficially acknowledged; very many theories in psychology, anthropology, sociology, history, and philosophy do pay lip service to it, but quickly resort to describing states, conditions, and patterns, or listing properties, in effect making very little of their dynamics, interplay, or responsiveness to change (see Section 2.1 below on the ubiquity of change). Climate change, tragic though it may be, provides an instructive case, which may compel us to face up to the fact that persons and societies can, must, and will undergo significant changes.

1.3 Wellbeing, Motivation, and Action

In Chapter 5, I survey the scientific literature on factors that support or inhibit behavioral change. It turns out that knowledge of climate change and its consequences is unlikely to make a positive difference. Nor are pro-environmental attitudes significantly correlated with pro-environmental action (Kollmuss & Agyeman, 2002; Moser & Kleinhückelkotten, 2018). This is a main reason for exploring the role of wellbeing in behavioral change. The relationship between felt and imagined wellbeing, motivation and readiness and ability to act is complex and still subject of debate. But it is beyond doubt that rewarding experiences (Wood, 2019) and prospects for future well- or illbeing—including that of other people—are important determinants of human behavior. This is not just an implication of classical behaviorist or belief-desire psychology, but also acknowledged by more "humanistic" approaches to motivation, like self-efficacy (Bandura, 1977) and self-determination theory (Ryan & Deci, 2017). It is explicitly recognized by the UN in their sober recommendations for how to foster sustainable lifestyles (UN, 2023b). It is also clear that sufficient personal wellbeing is a precondition for, or at least conducive to, adequate performance and engagement in benevolent (for example, pro-environmental) action, a relationship already noted by Kant (2012, 399 (BA12)) and confirmed by contemporary psychology (Isham et al., 2021, Welsch et al., 2020). In other words, apart from being a fundamental value (i.e., normative factor) in climate change mitigation and adaption, wellbeing is also instrumental to it.

In a pioneering discussion on the relationship between wellbeing and environmental responsibility, Bengt Brülde has raised doubts about the practical relevance of wellbeing theory. He worries that people can only be convinced that genuine happiness or wellbeing is more important than materially high standard of living if they endorse a traditional theory of wellbeing which gives priority to the former (Brülde, 2015, 93). But this is not how we should think about the relevance of wellbeing to the environmental policies. Theoretical knowledge about wellbeing is probably no more efficient than knowledge about climate change and its consequences when it comes to changing behavior. More important is how people perceive their situation and the affordances and obstacles for wellbeing. This also means that theoretical knowledge about wellbeing can indeed be practically relevant, but in a less direct way than what Brülde considers: not by motivating people to change their behavior or support environmental policies, but by guiding policymakers, innovators, behavioral designers, communicators, and teachers in their efforts to spur climate action. We must know wellbeing in order to work with it in a competent manner.

1.4 Views on the Relationship between Climate Change, Climate Action, and Wellbeing

To further set the stage, I will present a taxonomy of the different views on the relationship between climate change and wellbeing that have been voiced in current debates. Apart from providing an overview that can serve as a frame of reference for the subsequent discussion, it underscores my point that the understanding of wellbeing matters for how to understand and deal with climate change.

(1) *Only technological development can save us.* It is often assumed that only a technological breakthrough, which allows for continued, albeit "greener," growth (Jacobs, 2013) and consumption, can save us. Otherwise, the reduction in consumption and other restrictions required by green

transition—together with the cost of adapting to the climate changes that will, in any case, also happen—will impose an impossible burden on people. In the absence of technological breakthroughs and green growth, either climate mitigation action will be stalled due to the hardship and resistance it will cause, making climate change even faster and more harmful. Or else we might somehow manage to coerce ourselves into doing what we really do not like, but then look into a bleak future, having to sacrifice comfort and wellbeing to avoid catastrophe. This view is widely held by politicians and economists, but also by some climate scientists (Bowen, 2019; OECD, 2023).

(2) *Wellbeing must be sacrificed for moral reasons.* Some, among them several philosophers, agree that green transition will require significant sacrifices in terms of wellbeing from most present-day and near-future persons. But they consider it wishful thinking to believe that technological progress will come to the rescue. They take it that we have an overriding *moral* obligation to do whatever is required to keep climate change within limits. Morality trumps concern for personal wellbeing and can and should act as a sufficient motivating factor for making the necessary changes, unpleasant though they may be. This view has been explicitly defended by Brülde (2015) (and is implicit in Gardiner (2006) and Jamieson (2014)). It also seems to be endorsed by Mulgan, who argues that the prospect of a future "broken world" calls for a new (or reinterpreted) ethics that requires us to make significant sacrifices for ourselves (Mulgan, 2015).

(3) *Wellbeing can be maintained but must be understood objectively.* A third group of intellectuals agrees with the second that climate change does require us to curb consumption and alter our habits and behavior significantly. They likewise assume that this will seem and feel bad to many. But because they defend an *objective* conception of wellbeing, they do not consider this to be *really* bad. On the contrary, they assume that the change to a simpler lifestyle required by climate change can eventually be *good* for us, bringing us closer to the ideal life as this has been described by sages, philosophers, and folk wisdom down through the ages. Sometimes this view is presented with a moralizing undertone, implying that it is also good for people (because deserved, or purifying or edifying) to suffer somewhat while adopting the new and objectively more appropriate behavior. Climate change not only forces us to live better, but also teaches us a lesson, confronting us with the wrongness of our current ways of

living and engendering appropriate feelings of self-contempt. This view (though not necessarily with the moralizing aspect) has been defended by O'Neill (2017) and Rosa (2017). It is commonly associated with the degrowth discourse, which emphasizes that degrowth can maintain or even improve wellbeing (Alexander, 2011; Büchs & Koch, 2019; Schneider et al., 2011), though some proponents of degrowth also tend to support position 4) (see below). The view that sustainable living can meet objective standards of quality of life is perhaps most characteristic of the needs approach to human development (Guillén-Royo, 2018), but also associated with the capabilities approach of Sen (2000) and Nussbaum (2009).

(4) *Sustainability is compatible with subjective wellbeing.* Some agree with proponents of (2) and (3) that green transition does require us to curb consumption and alter our habits and behavior significantly, but do not consider it necessary to embrace an objective view of wellbeing. They argue that our current consumption patterns and other aspects of our present way of life do not really satisfy and are partly experienced as detrimental to wellbeing. We should therefore be likely, or at least able, to experience both the green transition and the new forms of life it will lead to as a change to the better—in terms of *subjective*, that is, felt or perceived or reported, wellbeing. The most detailed and sustained defense of such a view has been given by Kate Soper (2020), who stresses the implicit discontent with present consumer culture very strongly and cherishes the prospect of "new hedonism." In analytic practical philosophy, Peter Railton has argued that a closer look at the factors that are conducive to subjective wellbeing gives reason for optimism about the chances of maintaining such wellbeing (Railton, 2021).

Similar, albeit less philosophically developed, views can be found in many recent criticisms of contemporary society. A review by Kasser suggests that sustainable behavior is not just compatible with wellbeing, but that the two are mutually supportive (Kasser, 2017). Some support the position more indirectly, highlighting the objectively damaging consequences of growth, capitalism, and consumer culture to social capital, human relationships and job security (Douthwaite, 1993), or mental health (Wang & Tapia Granados, 2019), which are likely to take a toll on subjective wellbeing (Kasser, 2002), even though the authors do always not frame their discussion in terms of the latter.

(5) *Agnosticism: We do not know and cannot know.* A fifth position in the debate has seldom been explicitly defended in debates about climate change but is supported by more general work on human wellbeing and future life scenarios. (It may also be implicit in the very cautious assessment made by the IPCC.) It is the view that because of the general difficulties of forecasting the future, particular complexities and contingencies pertaining to climate change and green transition, but also the elusive and malleable nature of wellbeing, we cannot tell in advance what climate change or mitigation actions will mean to our future wellbeing. Support for the agnostic position comes from psychological research showing that humans are notoriously poor at predicting what will make them feel happy or unhappy (Gilbert, 2006). For example, people with serious disabilities tend to report much higher levels of wellbeing than people without disabilities consider possible (Albrecht & Devlieger, 1999). Similarly, recent work in philosophy has highlighted the phenomenon known as transformative experience (Paul, 2014): Some choices or activities engender new and life-changing experiences, to such an extent that we can know very little about our subjective futures. Even if we can reliably imagine the new conditions we will have to live under, we cannot know in advance how it will be like to experience them.

1.5 Why Not Gamble on Green Growth?

I shall discuss the strength and weaknesses of the different positions just outlined in Chapter 3. My sympathy lies mainly with (4) and (5). But my approach, also to the underlying theories of wellbeing, is ecumenical and pragmatic. All the positions, and most, if not all, dominant approaches to wellbeing are onto something and support similar conclusions in many cases. However, I think we can set (1) aside immediately. I will take for granted that technological innovation alone will not do—hence my assumption that significant behavioral change is needed. My justification for this is threefold. First, it is at best quite uncertain how fast and radical the development of new green technologies is going to be. Considering the urgency of responding to climate change, this obliges us to also pursue other courses of action, rather than gamble on a single risky option. Second, the very possibility of green growth (that is, of a fast decoupling between gross domestic product and greenhouse gas emissions) has been seriously called into question (Antal & Van Den Bergh, 2016).

Third and still more importantly, even if technological innovations should suddenly occur that will enable fast decoupling, it is very likely that the adoption of the new technologies will still require significant changes in behavior. Already, many technological solutions exist that could be used much more widely and effectively but have proven slow and difficult to implement. For example, it is not because there is a lack of designs that allow for durability, reuse, reassembly, or repair that product longevity (Jensen et al., 2021) and recycling are so slow to catch on. Quite to the contrary, companies and designers find it frustrating that consumers are slow to adopt the new solutions they offer (Gonzalez-Arcos et al., 2021). Wind energy is a tried and tested source of renewable energy that is in principle endorsed by almost everyone. But the expansion of wind energy has stalled in many countries, due to concerns for fairness and aesthetic values (García et al., 2016; Pasqualetti, 2011) and the so-called "not-in-my-backyard" syndrome (Jones & Richard Eiser, 2010).

It has also been argued that green growth, even if it should be possible, is likely to increase inequality and unemployment and so must be supported by radical social policies (O'Neill, 2020), which, again, will require a change in behavior, if not life form, of typical affluent Westerners. Generally, concerns for global justice and fair distribution of resources will probably make behavior change among parts of the global population inevitable, regardless of our technological advances. (It can also be argued that if it is possible to adopt a less costly lifestyle without impeding quality of life, then it will enable us to use the resources for other purposes, for example, supporting the wellbeing of people that are not so well off or promoting other values. Hence the rationale for considering this possibility can be formulated independently of concerns for green transition.)

To this can be added that the very notions of green growth and decoupling are ambiguous. Demonstrating the theoretical possibility of green growth is not the same as showing that a continuous increase in consumption is feasible. Gross domestic product (GDP) is a measure that is at once extremely broad and extremely narrow. It covers all sorts of monetary, but only monetary value. Hence a maintenance or raise in GDP is compatible with a reduction in actual consumption or the options available to citizens. If more is spent on environmental restoration, this will boost GDP, but surely not boost consumption. In such a case, there will be "green growth" on paper, but not in reality.

All this taken together should be more than sufficient, I think, to justify sidestepping the possibility that radical technological innovation will make substantial behavior change superfluous. Of course, this does not mean that we should not take the potential impact of technological innovations into account when trying to assess the quality of life in the future. It only means that it cannot make considerations of the possibility and consequences of new forms behavior irrelevant.

1.6 Why Not Just Politics?

Some readers may accept most of the points I have made so far. They may agree, for example, that knowledge, attitudes, and morality are far from sufficient to bring about the necessary changes quickly enough. But they may consider my approach too individualistic and for that reason perhaps also still not realistic enough. It may be thought that the most, perhaps the only, effective means for bringing about green transition and tackling the psychological obstacles are collective, and especially political, action. We need green laws and regulations. We need sanctions and incentives—taxing emissions and other activities that are harmful to the environment, perhaps banning those that are most harmful, perhaps subsidizing green technologies. It may be argued that in democratic societies, knowledge and attitudes can in fact bring about fundamental changes fairly immediately. For it is not necessary for the individual to develop new habits, which is known to be difficult and time-consuming (Wood, 2019). It is enough to vote in favor of green policies. If enough people do so, the politicians and existing mechanism of law enforcement will take care of the rest. Something like this has been argued by Wündisch, who criticizes Jamieson's green virtue-approach for being unrealistic and, at best, far too cumbersome to enable the necessary speedy transition, and suggests going for green votes instead (Wündisch, 2014).

However, some may think that democracy, at least in its present form, is also not efficient enough. It tends to get bogged down by the constant need to compromise and accommodate all sorts of interest and is biased toward the short term. On this view, the solution is still squarely within in the realm of politics, but the political system needs to be restructured. Perhaps we need procedures that allow for more swift and radical decision-making. (It is telling that when Wündisch speaks of green votes, he seems to be thinking of a referendum-like, directly democratic process, where a simple majority decision can cause a ban of increase in emissions, rather than the messier workings of typical representative democracy.) Perhaps we need to establish a "ministry for the future" (Robinson, 2020) with special tasks and powers. Perhaps we must resort to sabotage or terrorism (Malm, 2021).

Unfortunately, this is all too simple. Political action is necessary, but politics alone is not sufficient. If it were true that all we need are green votes, green transition would already have gone much further. Just as adoption of green technologies turns out to be surprisingly slow, both lawmaking and implementation of laws and regulations proceed slowly even in where there is wide support for green policies among the population. Apart from political support, there seems to be a need for compliance—or better, perhaps, concordance (a term that has gained popularity in the field of health communication and medical interventions, signifying an agreement and understanding achieved through negotiation and mutual adjustment, thus requiring actual involvement of the individuals who are expected to act in a certain way (Bissell et al., 2004; Britten, 2014)). This again points to the importance of the human interface in green transition. It is indeed a societal task and responsibility. But it depends on the ability and willingness of individuals to comply with what is politically decided.

It is common to speak of a gap between pro-environmental attitudes and actions (Kollmuss & Agyeman, 2002). When it comes to the relationship between political action (in the conventional, democratic sense of voting for lawmakers and enacting laws) and fighting climate change, I think there is a whole series of similar gaps which we should also mind: (a) Once elected, politicians do not set goals as ambitious as their supporters wished them to do. They fear that voters are not ready for shouldering the consequences of the opinions they, somewhat carelessly, endorsed on voting day, and go on to balance and blend their concern for climate action with other interests—partly out of a sense of responsibility ("lawmakers must see the big picture"), partly out of perceived political necessity. (b) Even when politicians do set ambitious goals—typically for long-term reduction in emissions, like in the Paris Agreement or national climate acts—they do not enact laws ambitious enough to meet the goals they have set. (c) Even when fairly ambitious laws are enacted, they are seldom implemented swiftly or completely. (d) Even when serious attempts are made to swiftly implement enacted laws, these are often met with local resistance, causing further delay and contributing to (a), making politicians think that they must proceed with caution and "get everybody on board," and so creating a negative feedback circuit.[3]

This still leaves the possibility that more radical political reform or action is needed. This is quite plausible. However, the problems of compliance and concordance cannot be so easily avoided. Democratically elected politicians may overestimate the risk of causing resistance among the population. But people do not like being coerced. The historical record of attempts at great societal changes from above is not flattering. We need not look to the extreme examples of the French Revolution or twentieth-century totalitarian regimes. Less comprehensive and seemingly more cautionary projects of changing societal practices through centralized planning and decrees have also proven difficult to carry out successfully, and often outright harmful (Scott, 1998). I am not saying that no positive examples of successful change through bold political leadership or grand design can be found. Rather, my point is, again, that the human interface cannot be ignored.[4] There is some truth to the notion that is important to get everyone on board. It has been characteristic of successful large-scale policy-driven societal transitions, like the welfare state expansion in Europe during the second half of the twentieth century or the rise of the Asian Tiger Economies, that there was something in it for the people it affected, also in the short term. Though the changes may also have been disruptive and challenging, they gave rise to perceptions of immediate improvement and reward. In cases where effective support has been marshalled for materially less rewarding policies, like war or self-imposed international isolation, this would hardly have been possible without incurring strong emotions and appealing to shared values.

Even if much could be accomplished through centralized decision-making and strict enactment of laws, it is beyond discussion that still *more* could be accomplished, and accomplished faster, by achieving acceptance, concordance, and identification, and ensuring that transitions are perceived as meaningful and rewarding. Since the extent and speed of climate action matters enormously, it should also be beyond discussion that political action, of whatever sort, must be complemented by studies and initiatives aiming at making green transition seem more rewarding and meaningful.

1.7 Wellbeing for Whom?

But from which perspective, it may be asked, should we assess the prospect for behavioral change and wellbeing? Climate change affects people in different parts of the world, and from different social classes, quite unequally. Climate action will also impact the lives of individuals and groups in widely varying ways, not least depending on how the burdens are distributed, but also due to differences in geographical location, available resources, etc.

I must admit that I will be mostly concerned with the situation of affluent inhabitants of the Global North. This is so for several reasons. It is the situation I am by far the most familiar with. But it is also objectively relevant. The carbon footprint of affluent people is much larger than that of those living under more modest conditions or in poverty. Hence the question of how to live with less consumption is most pertinent to them, and it is particularly important that they be able to change their behavior.

Some might find this focus unacceptable narrow and self-centered. It is the larger suffering and burdens of less privileged people that should occupy us, they might say. We ought to care less about our own wellbeing and more about those of others. That is correct. But it is also a fact that affluent people do care about their own wellbeing[5] and that this influences their actions and decisions.

That said, I think much of the subsequent discussion will be more broadly relevant. Going through transitions that are both challenging and hold promises for one's future life, either forced or voluntarily, is arguably part of the human condition. One can think of migration as an example a transformative activity that has many of the same features as radical change of life form in response to climate change. Besides, even though many inhabitants of the Global South will not necessarily have to curb their consumption or leisure activities, both climate change and climate change actions will nevertheless lead to significant changes in their way of life. The framework for assessing and working with wellbeing I develop in Chapters 3 and 4 is applicable to people living in less privileged circumstances. And although we shall not be much concerned with general climate ethics, it is a basic assumption of the book that we (that is, affluent inhabitants of the Global North) must change our way of life because of the harm it will otherwise cause to people in other parts of the world and future generations.

1.8 An Interdisciplinary, Philosophy-Driven Approach

The climate crisis is often described as a "wild" or particularly "wicked" problem—that is, a problem so complex and challenging that it is impossible to solve with conventional means (Rittel & Webber, 1973). The "wickedness" is supposed to stem from a lack of sufficient knowledge, which again depends on the systemic, multifactorial, and unpredictable nature of the problem, including the fact that it involves many different stakeholders and that it has both natural and socio-cultural aspects which are, moreover, almost inextricably intertwined.

It is widely thought that wicked problems call for radically interdisciplinary research (Sjöblom et al., 2024). However, much research on the climate crisis and green transition remains focused on specific aspects of the problems, in ways that make the significance of the results highly contingent on further factors that are not considered. The tendency noticed above to do work in climate ethics and politics without seriously considering the role of wellbeing, human psychology, or the challenges of personal and societal transition is just one case in point. The cautious, fragmentary, and surprisingly inconclusive approach of the IPCC report likewise shows an implicit recognition of the limitations of existing research. With this book, I try to take some steps toward a more integrated and genuinely interdisciplinary approach, while acknowledging that much more is needed, also in the specific (albeit complex and cross-linked) field of pro-environmental behavioral change.

My approach is both interdisciplinary and philosophical. I draw of findings and theories from especially psychology and social science more broadly, but also historical work and research on climate change and mitigation[6] and adaptation of a more natural-scientific or technical nature. Philosophy is present throughout. It functions as a knowledge broker that forges connections between disciplines. It is present in the form of a persistent normative perspective, as the focus is on the normative significance of findings and hypotheses, and the assessment of policies, strategies, and actions is done with an eye to their acceptability and wider existential and cultural consequences. Lastly, my attempt to maintain an unprejudiced, open-minded stance, challenging received views and taking very little for granted, is also typical of philosophy. Less typical of contemporary philosophy, but still very much in its spirit, I think, is an eclectic approach that combines not only theories and knowledge from different fields but also different styles of thinking—sometimes analytical, sometimes more synthetic and perhaps slightly speculative, sometimes even taking clues from literary fiction. A wicked problem like responding to climate change seems to call also for this.

The relationship between philosophy, psychology, and empirical science, however, is contested and complex (Klausen, 2022). It is beyond the scope of this book to go into a discussion about the nature and methods of philosophy.

I can only lay my cards on the table. I will, for the most part, maintain a fairly traditional view of philosophy as an investigation of the most general and fundamental concepts and states of affairs, which need not always be affected by what people think about them or what seems to be empirically the case. I think the value of this can be appreciated independently of any strong metaphilosophical commitments, even by people who are—like myself—interested in bringing philosophy and empirical science in closer contact, and not impressed with traditional claims on behalf of "pure" philosophy or conceptual analysis. The detached philosophical stance can help secure the necessary open-mindedness and flexibility, acknowledging that things and situations can vary widely, which is also an empirical fact. That a certain factor seems to be strongly correlated with wellbeing is no guarantee that it is the cause of it; nor can it be ruled that different factors would be more relevant under other conditions. I have difficulties seeing how a completely solitary life could be completely satisfactory, but it should not be ruled out in advance. By refusing to take empirical findings as proofs of how things necessarily are, philosophy contributes to ensuring that the investigation remains imaginative.

However, I also want my approach to be realistic. Climate change is real, and—as I have already stressed repeatedly—only solutions that consider real barriers to and affordances for climate action will do. So I will assume that human beings and the world, and their potential for change, are more or less as we know them to be. Importantly, I will not distinguish very strongly between philosophical and psychological approaches to wellbeing, action, and behavior. I will not be much concerned with whether a certain property or element is, strictly speaking, *constitutive* of wellbeing (that is, an absolutely necessary condition for it, something without which it could not possibly exist, under any circumstance), or "merely" a very general and fundamental, practically more or less indispensable *causal* factor (though I shall sometimes consider *how* fundamental or indispensable a putative element in wellbeing really is, in keeping with my critical-philosophical approach).

Finally, a few words about the structure of the book. The systemic, interdisciplinary, and cross-linked nature of the problem makes it difficult to structure the investigation in a systematic manner. Everything depends on almost everything else. But here is how I will proceed: In Chapter 2 I identify and describe what will subsequently be the object of interpretation and evaluation: the changes that we face in the future, due to both climate change itself and the mitigation and adaptation measures that we take in response. This of course includes our best scientific knowledge of the likely impact of climate change and necessary and possible responses. But I also consider how we do and ought to think about change, and how far and how we can distinguish different degrees and kinds of change in living conditions. Chapter 3 surveys the field of wellbeing theory, discussing strengths and weakness of different theories, and making tentative suggestions as to what is the most appropriate way of understanding wellbeing in relation to the climate crisis. This chapter contains the most sustained and, in certain phases, complex and technical, philosophical discussion. Readers whose primary interests are its practical implications may skip parts of it. It is also central, however, because it contains my main arguments for moving beyond the popular needs approach to quality of life in the context of climate change, and for viewing sustainability as compatible with subjective wellbeing. It also considers different forms of the so-called hedonic adaptation, arguing that it should be viewed more positively than is usually the case in discussions about wellbeing and climate change.

Chapter 4 returns to the topic of change, considering whether and how changes in lifestyle, preferences, or personality brought about by the climate crisis could have a distinctive impact on human wellbeing. It considers what might be called the existential aspects of climate-change-induced behavior change. Chapter 5 surveys behavioral studies and other research on how humans may or may not be able to change their habits or adapt to new living conditions. It relates this to the classic philosophical ideal of self-cultivation, exploring how far this could be a key or at least contribute to bringing about the necessary changes. Chapter 6 discusses whether different techniques and strategies for behavioral change are acceptable, mostly from an ethical point of view, but also with some regard to their aesthetic, rhetorical, and other features that might play a role in making them accepted or rejected or could hamper or facilitate their adoption. The chapter is thus both concerned with ethics and moral (etc.) psychology. Because a central issue is how far climate change measures are compatible with maintaining an autonomous or "liberal" lifestyle, it further addresses the question of how the quality of future life may be and what could be salvaged from current forms of living. Chapter 7 tries to answer the "so what?" question, showing how the lessons learned about the climate crisis, wellbeing and affordances and obstacles for behavioral change could be translated into climate action, both in the form of political initiatives and practices of individual and everyday life. A brief coda points to the more general, philosophical, and existential implications, beyond the case of climate change.

Notes

1 I am skeptical about the metaphysics involved in much contemporary work in the environmental humanities, including its normative implications. I think humans and animals *are* special in important respects and do deserve special concern, and I doubt that a sound basis can be found for ascribing moral rights to the inanimate nature. And since climate change is to a large extent human-made and must be combated by humans alone, putting humans at the center seems well justified. However, this is not something I can argue for in this book, nor does my argument depend on it. The important point is that regardless of their theoretical credentials, non-anthropocentric attitudes are not likely to be easily inculcated nor effective at bringing about actual change in people's behavior.

2 Gardiner does, however, acknowledge the need for a distinctive "ethics of transition," that articulates how we might proceed ethically "from … deeply constrained … social realities." But his actual attempt at formulating such an ethics consists solely in specifying further obligations.

3 Although in the context of reporting a case that eventually gives reason for optimism—the adoption of traffic congestion fees in the city of Stockholm—transportation director Jonas Eliasson has noticed that "the closer you get to implementation, the more drawbacks stand out," and aptly described the implementation phase as a "valley of political death" that one has to fight one's way through (Meyer, 2017; see also Ch. 7).

4 A similar argument has been made by Peter Singer as part of his advocacy of "a Darwinian Left," which takes human nature seriously (Singer, 2010).

5 I do not assume that affluent Westerners *only* care about their own wellbeing. Besides, assuming that they possess empathy, the perceived well- and illbeing of others, including others even in faraway places, as far as their situation is known and sufficiently vividly presented to them, will also influence their wellbeing.

6 I am aware we should, strictly speaking, distinguish climate change *abatement* and mitigation, but use the latter expression as a shorthand for both.

2

Life in the Future

We will have to change—but how much and how fast? Which changes will be necessary because of the climate, and which are required to reduce our greenhouse gas emissions to a sustainable level? This is something that we must at least be able to estimate if we are to assess our prospects for future wellbeing. The question is not only what we know about future living conditions. We must also consider the extent to which we do not, and perhaps cannot, know how they will be. But before surveying the available information about the impact of climate change and our possible responses to it, it can be helpful to consider still more generally how and how much human life seems to have changed and how these changes have been interpreted and perceived. We must try to look into the future; but as a preparation for this, we should also look to the past.

2.1 The Ubiquity of Change

Human life has always changed. At most times and places, change has been almost imperceptibly slow, with generation after generation living under very similar conditions. Consciousness of the depth and possibilities of historical change appears to be a relatively new phenomenon, judging from the blatant anachronism of medieval depictions of biblical events. Yet at some times and places, change has been rapid and thorough. The most striking example may be the immense technological, societal, and cultural changes that took place in Europe from the early modern period onwards, and which are associated with the scientific revolution, overseas exploration, and industrialization. But relatively speedy, and even dramatic, changes in living conditions have been much more widespread, and not restricted to what has come to be known as Western civilization. Throughout history, in all parts of the world, natural disasters, infectious diseases, climate change, war and occupation or breakdown of existing power structures have forced people to change their way of life within a single generation. Trade and cultural exchange have had a continuous impact on all but the most isolated of human cultures, to an extent that may not have been immediately noticeable, but with cumulative effects large enough to seem striking in retrospect. Especially the spread of new religions has taken place in an astonishingly rapid pace, as Buddhism, Christianity, and Islam swept large part of the world in a short span of time. While most people in many places may have plowed the fields in both the third and thirteenth centuries, the life and habits of a Roman peasant were, in other respects, very different from those of a medieval Welsh countryman.

It may be true that what appears to have been radically new ideas or practices were often adapted to existing ways of living, thereby making the transition seem less dramatic. Yet while this might say something about the conditions and perceptions of radical change, it does not mean that the changes in question were not indeed radical. Few today would deny that digital technology has altered our life pervasively in recent decades. It has done so precisely by becoming adapted to existing tasks and practices and supporting existing needs and goals—but to an extent that has made it recognizable as a force in own right, capable of shaping human behavior to a significant degree, for good but also for bad, and almost irreversibly.

Even the assumption that life in prehistorical times was almost uniform and unchanging may not stand up to scrutiny. New archeological and ethnographic research indicates that earlier human forms of life were much more diverse than has hitherto been recognized (Graeber, 2021). While we may be under the impression that it is only recently that human life has become diversified, fragmented, and fluid, as popular notions of modernity will have it, the overall trend in most of human history seems to have been one of ever-increasing diversification.[1] This can be seen from the welter of languages, ethnicities, and cultures that evolved after homo sapiens migrated out of East African, a process that has only lately been reversed by movements like Latinization, Arabization, Anglization, and, more generally, globalization. If we are to believe the revisionist anthropologists, humans have from early on known and experimented with different social arrangements. While they may have lacked a consciousness of large-scale historical change, they were aware of the contingency of particular ways of living and of significant alternatives, and able to move back and forth between sedentary and nomadic lifestyles (Scott, 2017) and different power structures and ways of distributing resources (Graeber & Wengrow, 2021), a fact that seems to refute the Marxist assumption that particular modes of production necessitate particular political structures and ways of life.

2.2 Change for Better or Worse?

We know, then, that humans can change their way of life, both profoundly and rapidly. But the only thing we seem to know for sure is that they do so when forced to, or that it simply happens. We do not know to what extent humans are able to change their way of life because they want to or think that it would be good for themselves and others they care for. That some crude forms of historical determinism can probably be dismissed does not rule out that human behavior is shaped by other, more subtle factors equally outside human control. Nor can we be sure that change, even of the apparently voluntary kind, is generally good for people. Many changes have arguably been to the worse. Almost all major inventions or putative civilization leaps in human history have been criticized for worsening the human lot: individual property (Rousseau and the Marxists), sedentary civilization (the new anthropologists, spearheaded by (Sahlins, 1972)), division of labor (Thoreau in *Walden*), industrialization (Henry George), market economy (again, Rousseau and the Marxists), digital technology, and so on. The Western literary tradition begins with laments of paradise lost, as Hesiod in *Days and Works* regrets being born in the Iron Age, marked by selfishness, fraud, and oppression. Though less attracted than Hesiod to the simplicity of rural life, Plato and Aristotle were similarly concerned about what they saw as a loss of aristocratic or gentlemanly values in the expanding Greek city states.

Even optimists about human history have tended to see much, if not most of it as part of a necessary sacrifice on the road to a more acceptable current or future state. The notion of a perennial tradeoff between gains and losses is extremely widespread, as even defenders of modern civilization or progress tend to see it as founded on compromises and limitations. In *Civilization and Its Discontents*, Freud speculated that humans have only managed to protect themselves against nature and mutual violence at the price of fundamental guilt and frustration, brought about by the need for self-restraint. Horkheimer and Adorno noticed in *Dialectics of Enlightenment* that "civilization's way was that of obedience and work," illustrating the modern human predicament with the myth of Odysseus only being able to enjoy the song of the sirens by letting himself be bound to the mast (Horkheimer and Adorno, 2002). Though not concerned with a similar deep and speculative diagnosis of modern civilization, Lukianoff and Haidt strike a similar note in *The Coddling of the American Mind*, speaking of "problems of progress," which they understand as "bad consequences produced by otherwise good social changes." They observe, for example, that "comfort and physical safety are boons to humanity, but … bring some costs, too," highlighting especially how improved circumstances lead to a reduced toleration of discomfort and risks (Lukianoff and Haidt, 2018). The notion of an inevitable tradeoff between costs and benefits has recently been trumped by a new wave of more outright pessimist assessments of contemporary human life. A notable example is Mark Greif's widely acclaimed *Against Everything*, in which he declares himself to "mistrust any authority that is happy with this world as it is" and attack a welter a things otherwise thought by many to be or having been conducive to a good life, like physical exercise, sexual liberation, enjoying food, the pursuit of pleasant experiences, and pop music (Greif, 2017).

Not only does it seem difficult to say if change has been, or can be, to the better overall. The standard by which to measure its goodness or badness is far from clear. The apparently strong case for people of earlier times, like hunter-gatherers, having led a life of greater affluence or wellbeing than their urban-dwelling or rural descendants turns on an understanding of wellbeing as being easily able to obtain the things one wants and needs (Graeber, 2021, 148). While this is indeed a popular and plausible notion, simply taking it for granted ignores the almost equally popular, and perhaps even more plausible, notion that humans can acquire higher wellbeing by developing, cultivating, and eventually satisfying new needs and desires, even though this may require them to curtail some of their freedom and sublimate some of their needs. The widespread criticism of present-day consumerism, which can be found in both religious and secular, even left-wing guises, typically goes so far as to assume that people seldom know what is good for them, and that satisfying their immediate wants and needs really makes them worse off. Hence the verdict that changes are to the good or the bad not only depends on a complex assessment of facts and weighing of costs and benefits. It also requires choosing between competing standards of what is fundamentally good for human beings. This further highlights the relevance of philosophical value theory, especially wellbeing theory, for seemingly empirical judgments. It also underscores the point that we should not be too quick to pass judgment on the prospects for acceptable personal or societal transitions. Neither the capacity for smooth change of life form nor its impact on the quality of life should be treated as a foregone conclusion.

2.3 Going through Changes

Assessing the impact of change is further complicated by the fact that is not just about comparing the value of different human arrangements obtaining at different times or places. The assessment must consider the significance of change itself to human life. It is a main thesis of this book that the value and disvalue of change, in particular its impact on human wellbeing and self-understanding, have not received sufficient attention in neither scholarly nor popular discourse. Yet this claim must be qualified from the outset. For the experience of change and its impact on human life and society has of course not gone unnoticed. It has been a popular subject of art and literature, history and personal recollections. Some might even argue that change, crises, revolutions, and other spectaculars events have received disproportionate attention, and that it is the more permanent factors that have been overlooked, a point perhaps most famously made by the French Annales school of historical writing (Braudel, 1982) but also known from the phenomenology, with its emphasis on the life-world and everydayness. And surely historical events, revolutions or significant societal developments, rites of passage or coming of age, uncertainties and fears induced by changing conditions, or they joys and triumphs of mastering change have been predominantly popular themes in art and popular discourse. Think of ancient epics like *Gilgamesh*, religious texts like *The Book of Revelation*, the *Bildungsroman*, and collective novels like *War and Peace* or *Middlemarch*. Goethe opens the Second Book of his seminal *Wilhelm Meister's Apprenticeship* by noticing that readers of narratives crave for development and action:

> Whoever strives in our sight with vehement force to reach an object, be it one that we praise or that we blame, may count on exciting an interest in our minds; but once the matter is decided, we turn our eyes away from him: whatever once lies finished and done, can no longer at all fix our attention.
>
> (Goethe, 2011, Vol. 1, Book 2, Ch. 1)

Even novels like *Ulysses* and *Mrs. Dalloway*, which are, at least the surface, confined to recording the—often humdrum—events of single day, depict them like an epic a journey and aim at capturing the flux of life.

All this must of course be granted, and I shall myself make use of the considerably body of depictions and studies of change and transition in art and science. My point is that when it comes to assessing situations and arrangements or justifying individual and collective policies, change is often either ignored or considered all too hastily and simplistic. In the quotation from *Wilhelm Meister*, even Goethe, who is otherwise unusually attentive to the qualities of development and transition, assumes that praise and blame go to the goals of striving, even though the latter may be more interesting. In the introduction, I noticed a similar tendency in wellbeing theory to effectively sideline the issue of change. It has arguably been ignored much more widely. Regardless of what else one might think about Robert Nozick's political philosophy, his criticism of the so-called end-state or patterned, as opposed to "historical," theories of distributive justice (Nozick, 1974) does seem to have a point, inasmuch as the former tend to ignore the relevant question about how to bring about the favored end-state or pattern in an acceptable way. And contrary to what might be assumed, even so-called existential philosophy has not been much concerned with the dynamic or process-related aspects of existence, apart from enumerating them and insisting on their importance. Heidegger's *Being and Time* contains an impressive taxonomy of constituents and modes of existence, couched in a language that is intended to be dynamic, as well non-objectifying, as well an abstract defense of the fundamental role of temporality. But it tells us very little about how the existential machinery can be brought to work, how life can be tackled, or how it unfolds. Only in very recent years, a new wave of analytic existential philosophy has emerged, which is more attentive to the problems of living and leading a life and the transformations it involves, and to which I am heavily indebted (Calhoun, 2018; Lambert & Schwenkler, 2020; Paul, 2014).

I am also not claiming that people are generally ignorant of change—quite to the contrary. Hedonic psychology has, at least to some extent, acknowledged the importance of perceived changes, and tends to confirm Goethe's observation: we are more interested in how things are going than in how they are. This tendency may be partly exaggerated, perhaps also somewhat culture-dependent (as Western culture is arguably characterized by a linear, progressive understanding of time and by giving particular emphasis to the radically new (Allik & McCrae, 2002; Ip, Chen, & Chiu, 2006), and to variation (Kim & Drolet, 2003)). Yet it is a fact about how a large part of humanity does think and tend to perceive the world, and a tendency which is spreading rapidly together with other Western values (Leung & Morris, 2010).

That changes matter a lot to people is further demonstrated by psychological findings indicating that their willingness to forego benefits in the present to ensure larger benefits in the future is influenced by the degree of connection they perceive between their present and future self (Urminsky, 2017). In other words, the prospect of discontinuity and loss of identity strengthens their preference for immediate gratification. This seems clearly relevant to climate change action. It indicates that a perception that changes in living conditions will not substantially alter oneself or what one cares about can foster support and compliance, whereas uncertainty about whether oneself and the world as one knows it will be preserved can lead to resistance.

Finally, it should be noted that in sharp contrast to the lack of focus on change and transition in wellbeing theory and climate ethics, the UN explicitly frames sustainability as a question of sustainable—and acceptable—*development*, indicating an awareness that perhaps the most important aspect of sustainability has do with the way it is pursued (UN, 2023). It is not that the importance of change has not been noted. It is rather that normative theory should face up to its ubiquity and central role in people's lives.

2.4 Quantity and Quality of Change

Asking how far humans are capable of change and what substantial change will mean for them presupposes that we can somehow quantify change—that we can say when an individual of a group has changed substantially. This, however, is also far from simple. For one thing, human life comprises so many different domains and aspects that it is massively *multidimensional* and cannot be assessed according to any single parameter. One might doubt whether a *scalar comparison* of different forms or ways of life is at all possible. Which of the two changes is the most substantial (or "biggest"): a situation in which the material culture—the tools, means of production, built environment, transportation system, communication technology, etc.—remains unaltered, but people adopt a new worldview (like in cases of rapid spread of religion or education and enlightenment), or a situation where the worldview stays the same, but technological innovation changes the material culture thoroughly? We would need to assign different weights to different kinds or manifestations of change to answer such a question, but it seems unlikely that this can be done in a nonarbitrary way.

A further complication has to do with the way changes are intertwined. Changes have an impact on the possibility or necessity of other changes. Conservatives like to say that it can be necessary *to change in order to conserve*, and they are no doubt right about this. In a way, this whole book is premised on it, assuming that we must change in order to maintain wellbeing. In the case of climate action, swift mitigation actions can make behavioral change forced by climate change or draconic future measures less acute. Or, to use a more concrete example: a change in consumption pattern, like an increase in the demand for reusable products or electric cars, can lead to lower production costs, reduced prices, and so further increase in the acquisition and use of environmentally friendly products. It may also spur research and innovation activities that could make such products still more efficient and/or affordable. With my rejection of green growth scenarios (see Section 1.4), I have ruled out that such feedback loops will in themselves be sufficient for driving the green transition. But they are likely enough, and my point is rather that their unpredictability—both when it comes to the initial impetus, the force of the "first movers," and the force of the subsequent feedback and cascade effects—makes it difficult to predict how our future living conditions will be.

Moreover, as Popper famously argued in his criticism of historicism, the course of history depends strongly on the growth of human knowledge, which is also radically unpredictable (Popper, 1957), This brings us back to possibility of unforeseen technological advances, but not only that: advances in our knowledge of human and social affairs may likewise change the situation we will be facing in the future, and so our need for certain other future actions.

I will try to cope with these complexities by combining an openness (or pluralism) toward the future with centering my investigation on what seems to be now the most likely scenario. I will maintain that the situation to which we must respond and adapt is so difficult to predict that we must be prepared for—and consider—very different ways in which it can play out. This is in keeping with the agnosticism described in Section 1.4, which I partly, but not wholly, endorse. For I think that are we are nevertheless able to make estimates that are plausible enough to guide our prudential considerations. It would not be wise to act as if we were completely in the dark about what the future might bring or how it will be affected by our current actions. I will center my investigation on a not unlikely scenario where relatively serious and efficient mitigation actions have kept the increase in temperature at bay, but not so much as to avoid considerable climate change, probably resulting in an increase in temperature well above the 2°C before the end of this century. That is, I will assume that we face a combination of threats to wellbeing caused by climate changes and climate actions in tandem. I will assume that we are going to live in a world that is not (yet) completely *broken* (as imagined by Mulgan, 2015) but subject to considerable strain—a world in which satisfactory life is still possible for most of its population, but in which special efforts are needed to achieve and maintain it.

As for the difficulties related to quantifying change, it should be noticed that we do seem to have a good intuitive grasp of the relative magnitude of at least certain kinds of change. We regularly identify historical periods like the sixteenth century, the time of the French Revolution or the First World War, or the 1960s, as particular "times of upheaval" and turbulence. We likewise assume that adolescence, getting into a new romantic relationship, beginning a new job or especially a whole new career, migration, divorce, retirement, and serious illness all count as significant life changes. We moreover intuitively tend to hold some factors constant or consider certain changes less significant. That people gradually grew older, or that technology and societal organization change incrementally is perceived as the normal, expected situation, and so almost not a change at all. We seem to intuitively apply a kind of Aristotelean substance-attribute ontology to cases of personal and societal change: some changes are perceived as superficial, pertaining only to the "accidental" properties of peoples or societies, and for that reason not as substantial—that is, not as expressing a change in the persons or societies *themselves*. Others are perceived as substantial, pertaining to the "core" properties of people (Moulouki et al., 2020) and societies.

Yet despite these robust tendencies to think about and even measure change in determinate ways, it must be granted that judgments about the magnitude of change are if not arbitrary, then at least influenced by particular interests and standpoints. There can be no question of measuring or ranking changes in a neutral way, prior to normatively assessing them. For what we wish to assess, the measure or "currency" of change, is its significance to human life, which is a normative and standpoint-relative property.[2] Transition into a condition of life in which there are limited opportunities for living an outwardly religious life may have seemed like a radical change to people of earlier times and may still seem so to people from some cultures, but it will hardly be seen as significant by typical modern-day Westerners. Judgments about the radicality and significance of changes in living conditions are, and must be, made from one's present perspective and situation, not from the point of view of the universe.

For this reason, we should be wary of saying that a transition to a much simpler, less consumption- and career-oriented life is no big deal, because the current consumer lifestyle is an aberration or anomaly, nevertheless, so we will just be returning to normal. Even if a form of life has been adopted quite recently, it can still have been adopted quite thoroughly and function as a normality standard from which significant departures are perceived as radical. I am not saying that we should work with an extremely and exclusively subjective notion of change. We need more and less subjective and objective notions to conceptualize different aspects of change. And surely there can be a point in saying: "It may seem to you that this has been a quite radical change, but look, it has only superficially been so; everything that matters more fundamentally, all the basic features of your life, are left mostly intact." This is also a typical means for hedonic adaption (for more on this, see Section 3.7). People often manage to restore wellbeing by recalibrating or "getting things into perspective" (Klausen et al., 2021), convincing themselves that seemingly stark events have not changed anything of deeper importance. My point is only that there is no neutral or standpoint-free measure of change—and that when it comes to assessing the impact of change on people's lives, wellbeing, and potential for action or behavior change, perceptions and standpoints should clearly be taken into consideration.

To get some traction on the issue, I will distinguish different *domains* and *dimensions* of change, assuming that it is somewhat easier to measure or rank the significance of changes *within* each of the domains and (especially) dimensions, while also attempting some tentative comparisons *between* domains and dimensions.

As for the *domains*, I will make a general distinction between properties or aspects of *persons* or *selves*, properties of *cultures* and societies (referring specifically to interpersonal relationships and patterns of organization), and properties of the *physical environment* and patterns of *physical behavior*. There is no discussion that these distinctions cannot be very clear-cut (as, for example, selves are arguably, at least partially or in some sense, constituted by interpersonal relationships). It is equally clear that the three domains are mutually dependent (as affordances for mobility or production arguably support, and maybe in some cases determine, modes of social organization or interpersonal relationships). All this should be granted, but the distinctions can be made and used meaningfully, nevertheless.

Now I will submit that the importance or significance of the three domains should be ranked as I have just listed them. Changes in persons or selves are, *ceteris paribus*, more significant than changes in cultures or societies, which are, *ceteris paribus*, more significant than changes in the physical environment and physical patterns of behavior. This may be said to reflect an anti-materialist bias. But it should be kept in mind that the domains are mutually dependent. It is possible, indeed likely, that the material conditions causally determine how things are in the other two domains. My proposed ranking (which, I repeat, expresses a normative and standpoint-relative assessment) merely entails that changes should count as more significant insofar as they are manifested in changes in personality, or subjective experience or interpersonal relationships.

When it comes to the *dimensions* within the personal domain, an overall distinction should be made between *experiences* and *personality traits* and *behavioral dispositions.* I take a person's *values, attitudes,* and *beliefs* to belong to the latter dimension, as I understand them as sets of dispositions to experience, think, or behave in specific ways (Schwitzgebel, 2022).

Within the subdimension of experience, we can again distinguish between on the one hand different *types* or *modes* of experience, for example visual and auditory, but also aesthetic or erotic experience, and on the other hand different *contents* of experience. This distinction is also far from clear-cut. Some might say that what distinguishes a reading experience from an erotic one is only their different objects or qualities (and some might perhaps even insist that one and the same experience can be both a reading *and* an erotic one). Still, going from seeing two brown cows to feeling cold does seem to be more of a difference in kind than going from seeing two brown cows to seeing four white clouds. This indicates a further ranking: a change in the *type* of experience is, *ceteris paribus*, a more substantial change than a change in experiential content. The *ceteris paribus* qualification is important here, however. There may be more and less central types of experience; and maybe a change in content of a central type should still count for more than a change between two more peripheral types of experience. A strong intensification of one experience—say, going from having very mild, barely noticeable back pain to feeling continuous, excruciating pain—may count as a more substantial experiential change than exchanging a musical for a culinary experience.

As for the personality traits, we have, fortunately, more than armchair intuitions to go by. A significant body of psychological research has revealed a pattern in the way people think about changes in personality. People's thinking about changes in a thing is generally guided by the principle that those of its features that participate in many causal relationships are most defining of it. Hence changes in such "causally central" features are perceived as most disruptive (Ahn et al., 2000). When it comes to personal continuity and change, people perceive changes in *morality* as the most disruptive, followed by changes in features related to *personality, preferences, experiences,* and *memories,* with the latter two being ascribed a relatively modest significance (Molouki et al.,

2020). However, the degree of perceived change is also found to depend on the *valence* of the change. Positive changes are generally not perceived as significant, whereas negative changes, for example a person going from being kind to cruel, are perceived as highly significant. There seems to be a default assumption that some kind of development is the natural state of affairs, and that people come to show what was, all along, their "true" self, when they show signs of moral improvement (Tobia, 2015; Strohminger, 2017).

Although it is noteworthy that change in experiences is not considered significant, it should also be noted that this probably relates primarily to the content of experiences (*what* is experienced, rather than *how*), and that the relatively low ranking is made from the third-person perspective. Changes in what *other* persons experience are not thought to change them very much as such. But this does not rule out that change or stability of one's own first-person experience does matter significantly to one's sense of personal continuity. I am not sure about how to rank the experiential and the personality trait dimensions, but will, for reasons of simplicity, treat the former as subordinate to the latter.

When it comes to the two other domains, we must again resort to intuitions and selective use of taxonomies made for other purposes. Most extant taxonomies of *life domains* cut across my divisions—they are also concerned with dimensions of *individual* life—but can be helpful in singling out the important dimensions, anyhow. (I include very close interpersonal relationships, like intimacy, romantic relationships, marriage, friendship, and even community (in a broad sense) to the personal domain. This accords with the philosophical tradition, which also typically treats love, friendship, and community as elements in a good personal life.)

First, it can be noted that *leisure* and *work* stand out as particularly important and salient subdomains of social life (see, e.g., Headey and Wearing (1992)); Vanderweele (2017)). Second, according to textbook sociological theory, the major components of *social structure* are *statuses, roles, social networks, groups and organizations, social institutions,* and society itself at large (University of Minnesota Libraries, 2016). A complementary, overlapping proposal takes social structure to be composed of *culture, social class, social status, roles, groups, and institutions* (Henslin, 2010, Ch. 4). Here, culture refers to a group's language, beliefs, values, behaviors, and gestures; and social class is seen as based on income, education, and occupational prestige. Statuses are stratified, as there are *master statuses* that cut across other positions that an individual may occupy

(Henslin, 2010). Social structure is also fundamentally characterized with respect to the degree of *integration* and *equality* and *inequality* (Wilterdink & Form, 2025).

Finally, as for the domain of physical environment and physical behavior, we can fairly straightforwardly point to *the natural* and the *built environment*, the means of *production* and means of *transportation* as important subdomains or dimensions.

Persons	**Culture and society**	**Physical environment and behavior**
Morality Personality Preferences Experiences Modes Contents Memories	Statuses (including social class; equality and inequality) Roles Social networks Groups Organizations and institutions Leisure Work	Natural environment Built environment Means and modes of production Means of transportation

This taxonomy is neither exhaustive nor exclusive. Often one and the same change will affect different domains or dimensions, either simultaneously, because of overlap in the categories, or through causal interaction between them. The taxonomy is admittedly highly arbitrary. But it reflects widespread and intuitive ways of dividing human life into fundamental categories. I think it is somewhat more than a checklist. But even if it is only that, then at least it provides with a tool for making assessments of the significance or magnitude of changes that would otherwise be wholly unsystematic, if not idiosyncratic

The domains and dimensions outlined must be applied to expected or possibly climate change-related activities and events: How much will this or that change in behavior or way of life affect this or that dimension? This will occupy us in the subsequent chapters. But let me here just initially consider a thought that will also occupy us later. For it may be tempting to think that the climate-related changes, at least those connected to mitigation activities, cannot be very large or fundamental—especially after surveyed what are arguably the basic domains and dimensions of life, most of which seem likely to remain unaffected (perhaps apart from the means of transportation, which are arguably not so central, anyhow). Morality, social roles and networks, etc., need not suffer, but might turn out to be strengthened by abandoning the consumerist lifestyle; giving up eating meat seems like a very little thing to do, and we have already seen—during the pandemic—that we have technological means for maintaining most of our social functions and connections with much less physical transportation.

I will eventually concur with this view, arguing that the changes necessary for climate change mitigation and adaption need not be very fundamental, even though they may be *extensive* and superficially disruptive. But I do not think this is something we can conclude immediately. First, the *experiential* dimension looms large as something the importance of which cannot be denied without question-begging, and it is likely to be significantly affected, even though the precise extent to which this will be the case is also an open and complex question.

Second, as noted above, changes are *nested* and *interrelated*. Physical changes can have symbolic or mental effects. The very experience of noticeable changes in ways of living, even if they can be deemed superficial, may itself be unpleasant and demotivating, and so make not so fundamental changes appear fundamental and count as such within a human life. There are plenty of everyday examples of small things—bad habits of various sorts, like using fillers in speech or picking one's nose—that should be easy to dispense with, also because they do not seem particularly rewarding. One of the shortcomings of the moralizing or idealizing views on pro-environmental behavior change is that they tend to depict it as no big deal, because it only concerns what is superficial, anyhow—thus underestimating the possible psychological or cultural centrality of things that seem trivial or superficial.

Third, there is the possibility of *cumulative effects.* Perhaps many small and, taken in isolation, insignificant chances can add up to something genuinely significant. This is what is assumed when it is estimated that adopting a sustainable lifestyle will lead to a decrease of wellbeing in the short run, because quite significant reductions in overall consumption will be required (Brülde, 2015).[3] It may be no big deal abstaining from having pets and eating little or no read meat, and not so big a deal switching from private to public transportation, and curbing flying, and turning the heating down by a couple of degrees, and buying less and reusing more clothes. But having to do all this within a short time span may be quite a deal and impact your wellbeing quite considerably. In short, we should not quickly dismiss the changes required for green transition as small or insignificant.

2.5 Living with Climate Change

Climate change has long been a fact. Its consequences can be felt also in Global North. The weather is becoming more extreme. There are more and more intense heatwaves, and draughts and floods are becoming more frequent and severe. The still stronger impact of climate change in the Global South is also felt indirectly in the north, as it is causing political instability and mass migration.

It can be safely assumed that a situation not just like this, but still more extreme, will obtain for at least the next hundred years or more, regardless of whether we are able to reduce greenhouse gas emissions significantly or not (if not, it will only become more extreme). What is difficult to predict is not so much the impact of the changing climate itself but the potential impact of adaptation measures. It is also difficult to predict how strongly the harms induced by climate change will be felt throughout the world, or whether some regions and population groups will remain shielded from the impacts, as is now very much the case. Here we should keep in mind that we are not directly concerned with climate ethics. It is not a question of whether it will be right if such a shielding persists, only of whether it is likely and what will be its consequences for wellbeing. For example, it is quite possible that shielding alone will not suffice to maintain current levels of wellbeing. Living on "islands" not directly affected by climate change might still impact negatively on the wellbeing of the islanders, just as living in societies with large differences in wealth and segregation of different social groups can affect the wellbeing of the well-to-do negatively (Buttrick, Heintzelman, & Oishi, 2017), and relative shielding does not preclude negative effects due to community-level interactions and linkages (Haines & Eby, 2019).

Some potential consequences of climate change for future wellbeing are very concrete and can be noted without any special consideration of normative or existential factors. This holds in particular for its influence on physical health. The IPCC points to numerous such effects, some due to exposure to heat, but also due to increased spread of various infectious diseases. But these are estimates of effects on the total world population. Apart from the fact that exposure to heat and especially the growth in infectious diseases will be much stronger in the Global South, climate change will affect the weaker groups within each population more severely, in some cases almost exclusively. Older adults are particularly vulnerable. By contrast, a now healthy and affluent person, especially if she is living in the Global North, will likely remain resilient to most of the direct climate change impacts on health listed by the ICPP (see the instructive map in Cissé et al., 2022, 1052). Again, this is not to say that it is morally unproblematic. Quite to the contrary, the extreme differences in risk and vulnerability further highlight the problem of global health inequality; insofar as this is a moral problem, climate change makes it more serious. It is only that these most concrete and manifest impacts of climate change on wellbeing will probably not affect the affluent inhabitants of the Global North. It may have a negative influence on their life expectancy, all other things being equal, and this will of course also be a reduction to their overall wellbeing. But all other things are not equal. The health effects of climate change will likely be counteracted by other factors that have generally led to a steady increase in life expectancy and will probably continue to do so (Crimmins, 2021).

However, as the potential indirectly negative consequences of shielding illustrate, climate change may have more subtle, but nevertheless significant ramifications. It may have *systemic effects* due to the interplay of different factors, the passing of unforeseen thresholds, and tipping points. (Note that I am not here thinking about tipping points in the physical *climate*, but of disturbances of social and existential equilibriums. Though it is almost impossible to estimate the likelihood of this, and still more difficult to estimate its eventual consequences for wellbeing, it is likely enough to deserve being taken into prudential consideration, as it provides a further reason for precaution, and so for supporting climate action, from a purely prudential perspective. For social tipping points in a more positive sense—the possibility that pro-environmental behavior suddenly goes from being rare and difficult to being the norm—see Sections 5.6 and 6.5.)

Now let us try to take stock of the influence of climate change of the domains and dimensions of human life. The physical environment will change quite significantly, though again, probably less so in the Global North. Coastal zones will be exposed to more and more severe floods. Their impact will depend on the extent of socioeconomic development in the vulnerable zones as well as the installment and reinforcement of flood defense structures (Vousdoukas et al., 2018). The coastline itself may change. If emission rates will be very high, this could trigger rapid ice sheet collapse, causing the sea level to raise to a level as much as 2 meters higher within the next century, and 3.7 meters higher around 2,150 (Lindsey, 2022). The flora will change, with conifers being exchanged for, e.g., chestnut and walnut trees (Experimentarium, 2021), though adaptation measures, like the introduction of more resilient tree species, could mitigate the changes significantly (Vacek et al., 2023). Northern Europe is generally expected to develop flora and agriculture similar to that which is currently associated with Southern Europe, as for example sunflower, lavender, and olive fields will become more common. Southern Europe itself will be more severely impacted. Assuming that farmland will have to be abandoned, this may significantly alter the landscape. This is not considered very likely however, as again, adaptation measures are available (European Environment Agency, 2020).

When it comes to the built environment, coastal areas will, of course, again be most strongly affected. Again, very much depends on both the regions since different regions have very different resources and starting points. For example, the US infrastructure is known to be already in bad shape and in urgent need of renewal. Cities may suffer badly from heatwaves, which can also stress the energy system. Cities may contain "heat islands," urbanized areas that experience temperatures 1–7° higher than outlying areas (UN Environmental Protection Agency, 2023). Wildfires will damage property in the American West and Southern Europe and affect air quality (Maxwell et al., 2018; Dupuy et al., 2020).

As for adapting the urban environment to climate change, refurbishing existing buildings is considered far superior, from a sustainability perspective, to demolishing them (Moncaster, 2021). Despite a widespread belief that cities will be more densely packed as ever before, there is evidence that densely built low-rise environments are more space and carbon efficient than high-rise buildings (Pomponi et al., 2021). Yet this is still consistent with the view "urban hubs" are superior to "urban sprawl" and single-family houses; they will likely take the form of localized communities that include mixed-use buildings and developments (Rutledge, 2020). This may lead to a relative loss of privacy and less living space per person, though on the other hand it will be more supportive of community life and also provide more varied activities and aesthetic experiences. Moreover, the effects of urban compactness may not be so significant in themselves, as they are mediated by a large range of factors (Chen, Jia, & Lau, 2008), and cities need not come to feel overcrowded. Hence assuming a sensible approach to urban renewal, there will indeed be significant changes to the built environment, but it is not likely to be to the worse. How far it will contribute to maintaining or even strengthening wellbeing depends, again, very much of the ability of the inhabitants of future cities to adapt themselves and exploit the wellbeing potential of new forms of urban living. As always, it also depends on the available resources and how far the capacity for carrying out the necessary transitions is hampered by systemic effects, like the consequences of climate crises in the Global South, and more generally, global and regional inequality.

This brings us back to the difficulties of prediction. I think my brief survey of the likely impact of climate on the physical environment in the Global North indicates that life may not, in general, become *that* bad under future conditions of global warming. But there are large unknowns. My relatively optimistic estimate presupposes that significant measures are taken to reduce emissions and, particularly, that much is done to carry out climate change adaptation measures.

Then there are two immaterial factors—in the domain of culture and society—that are also potential jokers which may upset the situation. First there is *money.* In my survey I have ignored the most frequently mentioned impact of climate change and climate change mitigation: the huge monetary costs they will incur. Climate change may "only" affect certain regions and production and commercial activities very severely. But this may still cause a considerable blow to the economy. And how far the modern—or, in this case, future—economy can absorb and compensate such a blow is very difficult to say in advance (though I tend to think about the capitalist economy as being quite adaptable and resilient). Likewise, the necessary adaptation measures may seem not only feasible, but so attractive that it is hard to imagine why they should not in fact be taken. What's not to like about a smart, community-supportive city structure? Well, it may prove so costly to realize that we should perhaps not assume it to become the typical form of urban environment, despite its obvious attraction. Moreover, climate change adaptation measures will have to compete with other tasks for financial and human resources. We know that preventive measures are seldom given highest priority, and it is likely that more will be spent on other societal challenges like the aging society, unemployment, or the development of rural areas, though it should be possible to combine climate change adaptation with meeting at least some of these challenges; urban hubs will be more friendly to older adults than urban sprawl, and adaptation measures will create jobs.

Second, life in the Global North may be affected not only through direct pressures from the more severely affected Global South, in the form of mass migration and political instability. We must also take the effects of *morality* into consideration. A situation with extreme segregation between different parts of the world is unlikely also because this would appear so unfair that in this case, morality would be probably become a real motivating and change-making factor. Concerns about extreme inequality may lead to changes in the global political economy, leaving less for the otherwise affluent and therefore resilient societies of the Global North to spend on adaptation measures benefiting their own populations. I do not think we should expect very large redistributions and resulting changes in the living conditions of inhabitants of the Global North, since they will hardly be politically feasible. Nor are the countries of the Global South likely to become powerful enough to pressure the rich countries into making substantial sacrifices. But we should reckon with a growing concern for global justice, with some shift in the global power balance, and with a pragmatically motivated interest, also among policymakers in the Global North, in carrying out climate change adaptation where climate change is most severe.

To sum up, there is a likely scenario in which the inhabitants of the Global North will not experience large changes in their way of life due to climate change itself, apart from changes caused by adaptation measures that may not have a large (or at least not negative) impact on their wellbeing.[4] But there is a risk that the monetary cost of the necessary adaptation measures will be too high or that concerns for global justice will remove the focus to other parts of the world. The latter is of course only a "risk" seen from the perspective of the affluent inhabitants of the Global North. I am not implying that there would be anything wrong with this, only that it could mean a more significant threat to the wellbeing of these affluent Northerners than the other changes considered here.

In any case, the most significant—but also the most unpredictable—changes will probably be in the domains of culture and society. They will be brought about by a confluence of direct and indirect impacts of climate change and our responses to it, for example in through intensified conflicts and migration, technological developments like AI and geopolitical changes, like increased multipolarity, with new states and regions gaining sway over the international agenda. Yet we should expect the unexpected, anyhow. That is how human life is. Change is ubiquitous and can and likely will come in unanticipated forms.

2.6 Living with Mitigation

My—relatively and tentatively—optimistic assessment of future changes due to climate change depends crucially on extensive and successful climate change mitigation efforts. These will also have a significant impact on various domains and dimensions of human life. Let us again start with the impact on the physical environment.

It has been proposed that a house or an apartment for three people should not exceed 1,600 square feet, and that we should eventually try to do with 450 square feet per person, also aiming at getting more people to share their living space (McLennan, 2014). A more recent study calculates that sustainable living would require the average floor space per person to be reduced from 754 square feet to only 215 square feet, similar to the average amount of space that was available in the United States during the 1950s (Cohen, 2021). This *is* a significant reduction. Saying that it will "only" mean a return to the living conditions of the very near and mostly positively remembered past would be missing the point. Life in the 1950s was very different from life as we know it today. It is remembered positively mostly because of the upward mobility, economic growth, and societal development characteristic of the period; this should not obscure the fact that most people had to do with very much less than today.

Looking at *transportation*, it seems likely that electric cars will come to dominate for some decades, without changing general transportation habits or perceptions significantly (with some possible exceptions considered below). In the longer term, private transport (apart from non-motorized forms, like bicycling) may be gradually give way to forms of public transport smarter and more attractive to the users than those known today, though the balance between private and public transport is difficult to predict. If public transport (again) becomes dominant, this will be a quite significant change, also in the domains of social and personal life. It should be noted, however, that a considerable part of the world's population already uses public transportation exclusively or predominantly, also in the Global North; think of cities like New York, Paris, or Tokyo. Hence it will rather be one familiar form of transportation coming to dominate another, just as certain more communal forms of urban living may come to dominate the suburban one. However, the subjective impact on those who will have to change, and who may perceive it also as a forced change in culture or a loss of personal freedom, should not be underestimated.

Transportation is of course also connected to leisure. Studies show, not surprisingly, that travel time and travel costs are central determinants of the choice of mode of transportation (Jowono et al., 2023), given a fixed goal (indicating that the vehicle of transportation as such is not so important, and that changes in this regard will not seem great). The choice of destination is a much more complex matter. It has been shown to be mediated by very many factors, including the relationship between valued activities and known features of the preferred destination (Moscardo et al., 1996) and experiences of joy, love, positive surprise, and satisfaction (Hosany & Gilbert, 2010).

The extent of and need for transportation will also depend on the future role of *digital communication.* We have already seen a reduction in travels and physical meetings in increased use of online meeting platforms, strongly prompted by the COVID-19 restrictions. It is likely that this tendency will continue, though online meetings, and online interpersonal relationships more generally, are widely seen as inferior to physical meetings and relationships both terms of wellbeing and the quality of outcomes and decisions. It should also be taken into account that digital technology is not carbon neutral. The large energy consumption caused by digital services has recently come into critical focus (Jones, 2018). Still, communicating online or engaging in virtual social activities will be relatively less energy consuming that getting together by traveling physically, and so it will probably come to play a larger role.

What about the impact on the *labor market* (an issue related to both the means of production and organizations and institutions)? Reduced consumption will require less production and so might lead to less need for work. However, this relationship may not be so simple, as more sustainable forms of production may also be me more labor-intensive. Yet assuming that less working hours will be needed, overall (also due to new technology), there is both the prospect of a rise in unemployment (as proponents of green growth are keen to emphasize) *and* the prospect of more leisure time, if the work becomes more evenly shared among the workforce (Vyas, 2022). There are economic models indicating that the necessary reduction in consumption will lead to an almost chronic recession, making unemployment and public debt skyrocket. However, adjusting the parameters, assuming that jobs can be shared more evenly and that new green investments will compensate for those lost due to reduced consumption, leads to a more optimistic forecast (Heinrich Böll-Stiftung & ZOE, 2022). Probably the effect of reduced consumption due to concerns for sustainability on average working hours, employment rate, etc., will be small, if not negligible, compared to that of technological development and changes in the way work is organized and distributed.

In some fields there will of course be more substantial changes, not least in agriculture and the fossil fuels sector. But they will affect relatively small groups. The net job shift due to green transition until 2030 is estimated at 10.3 million worldwide, with 3 million jobs lost but 13.3 million new jobs created (World

Economic Forum, 2022). As in the fields of housing and transportation, it might seem as if the green transition benefits the educated middle class and supports its manners and cultural preferences. The lost jobs may be low-skilled, and those who lose their jobs may not be qualified for or interested in the newly created jobs.

Now we should indeed be mindful that changes that seem to be, on the overall, not so radical or problematic, may nevertheless affect vulnerable groups strongly and negatively. Green transition could exacerbate social inequalities and deepen cultural divides; such tendencies are already visible today. They cannot be ignored just because we are concerned with the prudential rather than moral aspects of green transition and view it mostly from the perspective of affluent and educated people. For changes that directly affect only parts of the population, or are perceived rather than real, can nevertheless create frustration and social unrest and instability to an extent that does affect the wellbeing of citizens in general. Something like this can be seen in many countries today, where people who are feeling marginalized are able to make their mark on the social and political climate in ways that are perceived as disturbing. Social division generally takes its toll on wellbeing (Liu et al., 2014). Besides, the estimate cited above is only for the short term; in the longer term, a considerably larger number of people will be affected by the change to a greener economy.

Still, the problem should not be exaggerated. The job shift tendency is generally positive because it matches the growing relative demand for skilled jobs and supports social mobility. It is also wrong to think that low-income groups will in general be negatively affected and high-income groups benefited by the green economy. Farmers and others working in agriculture in the Global North are generally affluent. But they also belong to an already small and dwindling segment of the population. In high-income countries, only 3 percent of the workforce is currently employed in agriculture (Ruy, 2023). Hence it can hardly be said that green transition will cause a significant change in the labor market, since the number of people working in agriculture or mining, etc., has already been dwindling for a very long time. The shift from an agrarian or manufacturing to a service economy is very old news; the change in this direction is so familiar and expected that it is not something that should be perceived as a substantial change at all. Like with the change in working time, green transition just seems to feed into a more general trend.

When it comes to *food* habits, the most widely noticed consequence of green transition is that people will have to curb meat consumption, especially their consumption of red meat. It has been estimated that a sustainable consumption level would allow a person to consume 14 grams of beef, lamb, and pork and 29 grams of poultry per day, corresponding to 30 and 62 calories, respectively (Willett et al., 2019). In contrast, current average consumption in Denmark amounts to 44 kilos of beef, etc., and 8 kilos of poultry a year per person (Fagt et al., 2018), which translates into 113 and 21 grams a day. This is of course a quite considerable reduction. The amount of red meat to be consumed will be only 1/8 of the current level, whereas there will be no need to curb the consumption of poultry. The situation in the United States is still more extreme. Here, consumption habits have switched, so that poultry now accounts for a larger part. The average American consumes 68 kilos of beef and pork and 51 kilos of poultry per year, equaling 186 and 139 grams a day and requiring a reduction to $^1/_{13}$ and $^1/_4$ of the current level to achieve sustainability (Kuck & Schnitkey, 2021). Emerging economies are particularly interesting, because meat consumption is expected to rise here in the coming decades. In 2021, Chinese consumed on average 45 kilos of meat (of all sorts) per person; it is estimated that this figure will have risen to 52 kilos by 2029, equaling the current Danish consumption level.

All in all, the reduction in meat consumption necessary for achieving sustainability is quite considerable. Its subjective impact will depend strongly on improving and marketing meat substitutes. Although concerns for global justice and the association between increased meat consumption and perceived progress and social mobility may act to the contrary, the reduction should have a less noticeable impact in the emerging economies, where people have been used to consuming less meat. It is notable that most culinary traditions have until recently used meat in small proportions, sliced and shredded, used as topping, etc. Hence there is a large and well-established cultural repertoire that can be revived and developed and no need to break sharply with the past or abandon one's cultural heritage. Like in many other domains—transportation, forms of housing, etc.—the United States is an extreme and possibly more different case, as the required changes are larger here, and may clash more strongly with deeply rooted habits.

Food habits, and especially meat consumption, are widely assumed to have special symbolic and cultural value (Nungesser & Winter, 2021). We might expect that changes in eating habits will have a particular impact on *statuses* and *roles*, and also on persons' self-understanding. However, the associations between hierarchical positions in society and eating habits have already loosened to a considerable extent. Practices of food preparation and consumption no longer seem self-evident, and a modern, individualized subject of food choice has emerged (Coveney, 2006). Red meat has been associated both with the powerful and, in wake of the industrialization, the working class. It has been a social marker, but a quite flexible one. So it seems that we have here again a case of a societal development already being well under way, into which now feeds the concern for greenhouse gas emission. We have also here a further field in which groups that now only form a minority may feel encroached upon and forced to abandon their habits and preferences due to converging strands of modernization.

That food has to a certain extent lost its character as a clear social marker does not, however, mean that food consumption is no longer a central domain of human life. Quite to contrary. In line with observations already made by Simmel (1997, 130–5), meals have in part become a highly aestheticized activity. Yet Simmel was probably right in giving priority of the form of food consumption over its substance. It is not *what* you eat but *how* it is prepared, presented, and consumed which matters. The need to curb meat consumption will not reduce the general status and value associated with eating and food culture. The trend toward less meat and the trend toward emphasizing the aesthetic and social aspects of eating seem to go together quite well. The habitual, self-evident character of meals and food culture may not be preserved. But this is a change that happens independently of green transition.

A remaining worry concerns a possible loss of *richness* and *diversity* if meat consumption is drastically reduced (in Chapters 3 and 4, we shall study the general importance for wellbeing of these generic qualities). How far this can be avoided depends on how creatively low-meat dishes and diets can be prepared. Vegans are probably right that even the much more restricted vegan diet need not be boring, and we will still be allowed to consume meat in the modest proportions typical for many traditional dishes. Still, in the transition phase, it is likely that many will experience their diet as less varied and perhaps less exciting, especially when it comes to everyday food.

Clothing consumption is particularly interesting field. Clothing is the kind of commodity most strongly associated with consumerism. The sustainability impact of clothing manufacturing is particularly high, and the global stretch of its commodity chains creates special challenges. An introductory article claims that "though there is in general rarely an easy and quick fix in sustainability matters, it is even less so in the textile and clothing sector" (Boström & Micheletti, 2016). The urge to buy new clothes is the paradigm example of an apparent need for, and possible addiction to, consumption. It is almost trivial to remark that *shopping* is not primarily about satisfying a practical need for certain products, but rather an activity that is pleasurable in itself (thus the expression "leisure shopping"), and also expressive of a person's social role and identity (Rabbiosi, 2014). Hence a possible significant consequence of a change in clothing consumption could be a loss of leisure and of the pleasure and self-expression associated with browsing and bargain hunting.

A general abandonment of shopping activities would indeed constitute a very large change in behavior (not least in the domain of leisure). Some might see both the urge to shop and the pleasure it brings as superficial, but this is probably wrong. Even if you are not into browsing and buying clothes, you will likely enjoy and have difficulties resisting other kinds of browsing, which seems to speak to quite fundamental, probably evolutionary hard-wired, inclinations for collection, hoarding, and hunting. But we can envisage a production and consumer culture that is considerably more, perhaps completely, sustainable while still affording the pleasures of bargain-hunting, browsing through commodities, etc. Convincing proposals have been made for how to maintain a creative *fashion* culture while avoiding overconsumption (Fletcher & Tham, 2014; Fletcher, 2019). Like in the case of food culture, the trend toward sustainability might support our inclinations, in their more cultivated form, rather than curb them. Since I will be arguing against approaches to pro-environmental change marked by ascetism or a preference for austerity, this possibility is important to keep in mind.

As with other commodities, the transition to sustainable forms of clothing consumption depends on more *reuse* of products and materials. Enhancing reuse and reassembly and creating what is called a *circular economy* sounds like one of the easiest, least burdensome strategies for pro-environmental behavior change. But very likely it is not. Throwaway culture is convenient and central to contemporary everyday behavior; the make–take–dispose model is not just a mindset or superficial economic arrangement, but a deeply ingrained habit. Hence the transition to a circular economy must be considered a change of some magnitude, though, like in many other cases, we are able to envisage ways of implementing it that would make it seem much less demanding in terms of actual behavior change. There are already many practices and systems designed to make our preference for easy disposal compatible with reuse (convenient forms of waste sorting and waste collection, using reusable or compostable plates and cutlery, etc.).

What about the economy and the political system more generally? The form it will take will significantly impact not just the means and modes of productions, but also statuses, roles, organizations, and institutions, and arguably also our individual psychology. I will assume that capitalism in some form will survive; our basic economic system is not likely to change. I do agree with those who think that capitalism is not optimal for dealing with the present state of the world, and that both the climate crisis and growing global inequality will put it under increasing pressure. On the other hand, it is supported by very powerful interest groups who are not likely to give it up. And while capitalism may not be so extremely smart as it is often held to be, nor capable of integrating all legitimate concerns, it does seem flexible enough to be able to if not solve the problems it is facing, then at least adapt to the situation.

How capitalism will evolve under conditions of climate change and mitigations is difficult to say. If politicians remain reluctant to redistribute revenue and compensate low-income groups facing higher prices because of increased taxes, we may expect a further growth in inequality. This could cause feelings of social inferiority and unfairness, leading to social unrest and possibly also an intensified, stressful jockeying for position, as the impacted groups will be struggling to retain their social standing and certain positional consumer goods may acquire more, rather than less symbolic importance.

However, there is reason to believe that both the climate crisis and the growing concern for global inequality will curb some of the effects of capitalism, not prompting its overthrow, but at least curtailing it. Though politicians are still notoriously reluctant to introduce carbon taxation, there will almost inevitably be a change from taxing income to taxing harmful activities like greenhouse gas emission. This will also alter the social environment, through social feedback effects, making some positional goods so expensive that will lose their wider popularity. If luxury cars get more expensive, fewer will be bought, and this will make them less attractive to other, just as taxing cigarettes have made them less widely present and so reduced the risk of people becoming smokers through social contagion (Frank, 2020, 191ff.). The effects on consumer behavior caused by the taxation policies that will have to be enacted in order to effectively reduce emissions can be seen as an example of the positive power of limitations, the normative aspects of which we shall consider later (see Section 3.9).

These are some of the likely *concrete* changes in lifestyle and behavior required for successful climate change mitigation (though I have also briefly considered the symbolic impact of some of them). They may, on the whole, appear fairly trivial and existentially insignificant, though I think Brülde (2015) is right in assuming that they will be far from easy to make for many inhabitants of the Global North. When it comes to their impact on the deeper, social, and personal dimensions of life, it seems limited. They seem to concern only the specific contents of certain types of experiences and actions. Or, to put it differently, they seem to concern only the degree of satisfaction of certain of our values and preferences, and not these values and preferences as such. They do not make us into different *persons,* and they may not even be said to change our form of life.

I am inclined to eventually agree with this sentiment. Indeed, a main argument for the possibility of not just acceptable, but gratifying transitions is that the transitions are not likely to very deep or disruptive, so that most, if not all, that is needed to live a satisfactory life can be maintained. Still, as I have argued above, the cumulative, indirect, or symbolic effects of the changes described may still turn out to be large and might *seem* even larger, also depending on how they are brought about. Much depends on the extent to which certain goods or experiences can be easily *substituted* for others. This is also a complex question, both philosophically and practically. Research conducted by myself and colleagues on the role of alcohol in the well- and illbeing of older adults indicates, first, that alcohol consumption has a negative influence on older adults health and seem to be little needed and not strongly desired. It also shows that alcohol does have an important symbolic function and that not being permitted to drink alcohol is experienced as a serious limitation of the older adults' autonomy and possibilities for enjoyment. But it further shows that some kind of substitution or compensation is possible, like promoting other leisure and celebration practices or consumption of nonalcoholic beer and wine (Klausen et al., 2020). The climate-related changes we are concerned with here may be similar in that they seem objectively insignificant and even positive, may nevertheless be perceived as harmful intrusions into people's personal life, but can also be implemented more or less cleverly and respectfully.

In any case, a more definitive assessment of the impact on wellbeing of the expected changes presupposes a deeper understanding of what wellbeing is, what supports and what threatens it, and how it should be thought of and handled in the specific context of climate change and green transition. This is the topic of the next chapter.

Notes

1 A point aptly made by Wilhelm von Humboldt, who feared that creative diversification was coming to an end due to increasing cultural uniformity and saw this also as a threat to the development of a well-rounded personality (Humboldt, 1967).

2 It may be though that it is just hopeless and practically irrelevant to try to quantify change as such as it may be to try to quantify something like *health*. Hausman (2015) has argued convincingly that health is too multidimensional to allow for quantification, and that what we are interested in are rather its *impact on human wellbeing*. Yet in the case of *change*, I will try to work with a notion and "measure" of it that allows for an assessment of its magnitude or significance independently on its impact on wellbeing (so that it is possible, for example, that large and otherwise significant changes do have a large impact on wellbeing). This may be seen as roughly analogous to the idea, also endorsed by Hausman, that we need a neutral, naturalistic notion of health that for example allows that one and the same physical condition can affect the wellbeing of people in different circumstances differently.

3 Brülde speaks, more precisely, of a reduction in *happiness* (understood hedonistically, as defined by the balance of pleasure and displeasure), leaving it open whether that would also mean in a reduction in overall *wellbeing*. But if happiness is a central component of wellbeing, as I take I to be, it is not unreasonable to apply his estimate to this as well.

4 It will probably seem to some that I am belittling the impact of climate change. For example, David Wallace-Wells' much-read *The Uninhabitable Earth* (2019) seems to paint a much bleaker picture of the future. It might seem to flatly contradict some of my central claims, as Wallace-Wells argues emphatically that we should climate change to alter our ways of life quite significantly, that we should not assume its impact to be restricted to certain regions of the world, like coastal areas, and that wealth will not be a shield against the ravages of warming. Yet on closer inspection, the substantial (as opposed to rhetorical) differences are not large. Wallace-Wells admits that because of human adaptability and the expected mitigation efforts, it is unlikely that climate change will render the planet truly uninhabitable (2019, 15). I likewise emphasize that my relatively optimistic assessment assumes that extensive mitigation and adaptation measures are taken.

3

A Theory of Wellbeing for Green Transition

I have argued that understanding wellbeing is important to climate action and climate change policies. I shall now consider how it is best understood, looking at the existing theories, assessing their strengths and shortcomings and discussing how they may be combined, developed, and applied. As emphasized in the introduction, my approach is pragmatic and ecumenical. All the dominant theories of wellbeing are onto something, and the case of responding to climate change is better served by being constructive than by looking for marginal flaws. Still, it is not just theoretically, but practically important to use the theories in a critical and selective way. They are not equally well suited for contributing to solving the motivation and behavior change challenge. And despite their overall convergence, they do yield different results when it comes to assessing different policies and scenarios. This can be seen in the taxonomy of different views on the relationship between wellbeing and climate change (see Section 1.4). It is possible that we need different wellbeing theories to solve different tasks related to the climate crisis, for example, subjectivist, psychologically oriented theories to understand and support motivation and action and more objectivist theories to normatively assess the consequences of climate change and different responses to it.

My aim is to work out a *conception* of wellbeing. I will try to specify its constituents and go some way toward identifying causal factors that support wellbeing generally and robustly, that is, across different contexts.[1] But to make it clear what the discussion is about, in the first place, I shall assume a standard definition of the *concept* of wellbeing. I take it to be the notion of what is noninstrumentally good to for person (Crisp, 2021), of what makes life good for the person living that life—good for *her own* sake (Tiberius, 2015); compare (Darwall, 2002).

I shall, however, also consider theories that are not strictly speaking theories of wellbeing, notably the needs and the capabilities approach and theories of "the good life" in a wider sense, which may include moral standing. These theories do not attempt to explicate or flesh out the concept of wellbeing but propose to substitute it with a different normative notion.[2] It is often argued that especially when it comes to cases like climate change and international or intergenerational justice, such notions are superior to that of wellbeing in the narrow sense, that is, more appropriate to the evaluative task at hand. I will use the expression "quality of life" for the generic phenomenon all the theories under scrutiny are concerned with. I will argue both for a specific theory of wellbeing *and* for keeping to the notion of wellbeing (as the most relevant notion of quality of life) when discussing issues related to green transition and climate change.

3.1 A Plethora of Theories and a Provisional Assessment

It is common in philosophy to distinguish three main competing theories of wellbeing: hedonism, desire satisfaction, and objective list theories (Parfit, 1986; Fletcher, 2016). Hedonism takes wellbeing to consist in a favorable balance of positive over negative experiences. Desire satisfaction theories assume that a person enjoys wellbeing to the extent that she gets what she wants, that is, how far her desires are satisfied. Objective list theories see wellbeing as consisting rather in doing or achieving things that are good in themselves, independently of how they are valued or perceived by the person herself, though it can also accommodate having pleasurable experiences, as one of several different items on the list of fundamentally good things.

All three theory types come in many different versions. Quantitative hedonists recognize only one kind of pleasure or valuable experience, whereas qualitative hedonists distinguish different kinds of pleasure and assume that certain experiences (for example, enjoying a great artwork) may be more valuable than others (for example, enjoying a warm bath), despite being of lesser duration or intensity. Some take wellbeing to consist in a favorable balance of positive over negative emotional states (Haybron, 2008),[3] though these must arguably also be experiential states of a sort (Klausen, 2016). I shall treat this, in many ways attractive, emotional state theory as a sophisticated variety of hedonism. Preference satisfaction theories differ on the extent to which a person's wellbeing should be seen as depending on her actual preferences or rather something more idealized, for example, what she would prefer is she were fully informed. And objective list theories may work with shorter or longer lists, comprising of different items.

In the social sciences, three different constructs of wellbeing are standardly employed and increasingly combined. The most widely used is that of *life satisfaction*, understood as how a person judges her life to be, overall. It is often operationalized in life satisfaction surveys, asking people to rank their degree of life satisfaction on a scale on so to "self-report" it. Because life satisfaction is construed as a general *judgment*, it is also referred to as *cognitive wellbeing* (Weimann et al., 2015). By contrast, *affective wellbeing* refers to the frequency and intensity of emotions and moods. It is argued by some social scientists to be a more a valid construct, closer to "real" wellbeing and less sensitive to biases and arbitrary standards of judgment (Kahneman et al., 2004). But it is also widely acknowledged that it is difficult to register affective wellbeing reliably, since as a "within-person" construct, it is more elusively subjective. The most popular method for measuring it is experience sampling by means of diaries or electronic devices. A third construct, that of *eudaimonic* wellbeing, is now also increasingly being used. It is inspired by objective theories that take wellbeing to consist in factors beyond positive experiences or desire satisfaction, notably theories of human "flourishing" of which the ethics of Aristotle (2020) serve as the paradigmatic example.

It should be noted, however, that as constructs for use in empirical social science, all three notions are operationalized in ways that rely crucially on *self-reporting* and assume that the persons under study are themselves able to gauge and tell reliable how far they can be said to enjoy wellbeing. Experience sampling is in effect a kind of more narrowly focused "micro-survey": people are asked more often, and more exclusively, about their emotional states, but it is still up to them to judge and report. Likewise, eudaimonic wellbeing is usually taken to refer to people's *experiences* of whether and how they flourish, live virtuously or the like, and these are also registered by collecting answers to questionnaires (Waterman et al., 2008). Hence all three approaches are, in a way, subjective.

The philosophical and the social science notions of wellbeing do not match perfectly but overlap and exhibit similarities. Affective wellbeing has obvious affinity to hedonism since both are concerned with a favorable balance of positive over negatively valenced mental states. Affective wellbeing is, however, exclusively about emotions and moods, whereas philosophical hedonism focuses on experiences quite generally, like tasting something delicious, seeing something beautiful, etc. Life satisfaction approaches to wellbeing have some affinity to the desire satisfaction theory, as both make wellbeing dependent on the extent to which a person's life lives up to her subjectively determined standards.

It differs, however, by letting it up to the person herself to judge how her life is going, rather than (as on the desire satisfaction theory) measuring it by how it actually goes. As noted, the eudaimonic wellbeing approach draws inspiration from objectivist wellbeing theories, but likewise has a more subjectivist bend, seeing a person's experience of the qualities emphasized by such a theory, rather than their mere instantiation, as the decisive factor in wellbeing.

Both the philosophical and the social science approaches to wellbeing are relevant to the case of climate change. It is through the latter that we can gather information about the conditions that are conducive or detrimental to wellbeing. The philosophical theories, on the other hand, can function as a critical corrective to the apparent results of empirical wellbeing studies and give us some overall indications of which kinds and combinations of factors could be relevant in the first place.

As should already be clear from this brief exposition, the distinction between *subjective* and *objective* wellbeing is both important and far from clear-cut. It is not just that apparently "objective" notions of wellbeing are operationalized or interpreted in ways that gives them a more subjective flair. "Subjective" can mean different things. Desire satisfaction theories are subjective in the sense that the preferences that define what brings a person wellbeing are a matter of personal inclination or choice. They are objective, however, in the sense that it is usually things in the external world, which may even be beyond the person's own ken, that determine her wellbeing. If she desires to be loved but is merely deceived into believing that she is, her desire is not met, and her wellbeing is reduced accordingly, even though she is herself convinced that she is doing fine. Hedonism, on the other hand, is a subjective theory in the sense that it takes wellbeing to be *ontologically* subjective, consisting in experiential states. Yet it is objective in the sense that it is not *up to* the person herself to decide the criteria for her own wellbeing. Even if she does not herself prefer pleasurable experiences—maybe her religious beliefs make her despise them—lacking them will make her worse off. Subjective wellbeing in the sense of a positive mental state is often referred to as *happiness*. It is an open question to what extent happiness is constitutive of wellbeing. According to hedonism, happiness is all there is to it. Other theories view it as one component among others. In social science, life satisfaction is also used as a measure of happiness, though it is subjective in the sense that it depends on the person's own standards and judgment.

Now let us take stock of the theories and their attractions and weaknesses. The least attractive view seems to be the desire satisfaction theory. For it is beyond discussion that people can—and often in fact do—have preferences they had better not have fulfilled. They desire smoking, drugs, gambling, toxic partners, unrewarding jobs and positions, and many other things that either seem to detract from or at least do not contribute positively to their wellbeing. As Roger Crisp has argued, preferences and desires do not seem to be the kind of thing that could have fundamental value; their goodness seems to depend on the goodness of their objects (Crisp, 2006). This can also be seen from the fact that a few if any philosophers have actually held the view that the desire satisfaction *per se* has prudential value (Fletcher, 2021).[4] And when the view is expressed not in terms of actual preferences but in terms of what a person would choose if she were both rational and fully informed about the object of her desire (Fletcher, 2016), this indirectly confirms that it is the quality of the latter—the object of her desire—that really matters.

These are deep philosophical reasons for rejecting desire satisfaction theories. They are further supported by empirical research which shows that desiring and wanting something do not necessarily make a person happy when she successfully obtains it (Besser, 2020; Kahneman & Deaton, 2010). In the context of climate change and green transition, one may be particularly worried that people's present desires are not even a good indicator of what may be good for them in the long run. Their desires may be all too closely tied to contemporary consumerist lifestyle. It is imaginable that people could come to get significant wellbeing out of something for which they do not yet have any strong preference, like engaging in pro-environmental activities or enjoying nature.

That said, I do not think we should be too hard on the desire satisfaction view. Desire satisfaction does play an important *instrumental* role in wellbeing. Unsatisfied desires usually lead to frustration, which feels bad. Perceived desire satisfaction feels good. Something may be good to one even if one did not wish for it, but pleasures and rewards that go against one's preferences tend to feel less good and may give rise to negative feelings. Desires and preferences of a deeper and lasting sort are also part of what makes up a person's character. They motivate her actions and guide her decisions as to how to lead and sometimes reorient her life. People rightly care about the preferences they have or might come to acquire, often sticking to their existing ones and fearing that forming radically new ones will alienate them from themselves (Paul, 2014).

Moreover, while critics of desire satisfaction theories are right that people not always know what is best for themselves, in many if not most cases they probably do know it better than anyone else. They have developed preferences in response to their specific environments, experiences, and psychological dispositions; maybe they desire to be alone because they know that being among other people make them anxious or stressful. Hence preferences and desires are psychologically important elements in wellbeing that should be taken into account when we consider how to support it. Just saying that preferences do not determine what has value, or that people might be better off with something different from what they now prefer, is not sufficient, even though it is theoretically correct.

Something similar holds for the life satisfaction approach to wellbeing. Judgments of life satisfaction are likewise subject to influence from cultural norms and social comparisons, and they may not be made in a sufficiently well-informed and competent manner. People may not really make a balanced and comprehensive assessment of their whole life situation but tend to base their judgment on features that seem particularly salient to them at the moment or reflect their arbitrary interests. They may also lack cognitive access to aspects of their mental life that is important to their wellbeing, like their deeper emotional states (Besser, 2020; Haybron, 2008). But in many cases people do seem to have a pretty firm grasp of how things are going in their life, again probably more so than most other people. They know things that may seem unimportant from an objective point of view, but which nevertheless matters to them and so should play a role in the assessment of their overall quality of life.

Finally, it should be noted that preferences, desires, likings, and life satisfaction are different things and that each come in different forms, some of which may be more fundamental or prudentially relevant than others. A preference can be a very detached, "cool" attitude of approval or comparative evaluation, which does not play any deeper role in a person's mental life. For example, I prefer that all countries would have had right-side driving, but I really don't care much, and it is hard to see how this is related to my wellbeing. Desires and likings, on the other hand, have strong affective components. Life satisfaction can similarly both take the form of a judgment or disposition to judge one's life in a certain way and as a feeling of satisfaction with one's life.

Having set desire satisfaction theories to one side leaves us with hedonism and objective list theories as the remaining candidates for a theory of what wellbeing fundamentally is. I shall not try to make any definitive choice between them here. This is because I intend my approach to be inclusive (and because the best theory may turn out to be hybrid, a combination of different theories, which may even leave the preference satisfaction theory in the race, albeit as an outsider). But it is also because we are looking for the theory that can best shed light on wellbeing in relation to climate change and green transition. So instead of considering subjective versus objective notions of wellbeing in general, I will show, in the next section, how subjective wellbeing has been thought to be inadequate for this more specific purpose and use this as springboard for a more focused discussion.

Here, it should suffice to notice what has traditionally been regarded as the fundamental weaknesses of the two theories: Hedonism has been accused of identifying wellbeing with something that is too shallow, fleeting, and superficial (Haybron, 2008) and disregard the value of authenticity and contact with reality (as many take Nozick's "experience machine" thought experiment to have shown (Nozick, 1974, 1989)). Objective list theories, on the other hand, have been criticized for not being able to justify or explain which goods should be on the list and paternalistically stipulating that certain things just are good for everyone. They can be accused of disregarding the experiences and authority of the people whose wellbeing is up for discussion. Both hedonism and objective list theories can, moreover, be criticized for being end-state theories that ignore the path to wellbeing.

3.2 Problems with Subjective Wellbeing? Immeasurability, Growth Ideology, and Adaptation

As noted above, it is widely thought that the case of climate change demonstrates the inadequacy of subjective conceptions of wellbeing and provides special support for objective theories of quality of life. Four main reasons have been given for this. First, it has been argued that subjective wellbeing—be it in the form of experiences or emotions or satisfaction of personal preferences—is too *subtle* and *variable*. It can be influenced by very many factors, including very personal or erratic preferences, arbitrary standards of comparison, or fleeting moods and emotions. Thus, John O'Neill claims that "a central problem with subjective assessments of wellbeing is that different individuals in different contexts will appeal to different standards and expectations of what counts as a good life" (2017, 146). By contrast, so the argument goes, we need robust measures and criteria. And this need is best served by an objective conception that defines an acceptable standard of living in terms of basic goods.

Second, some see subjective wellbeing (sometimes wellbeing quite generally) as a concept that is problematically and perhaps inextricably wedded to ideas of *unlimited consumption and growth*, driven by self-interested individuals, and so incompatible with the concern for sustainability, which is a concern for balance and appropriateness, rather than for expansion (White, 2017; Rosa, 2017). The climate crisis requires us to remove the focus from what could make this or that individual or particular group feel better, to how we can ensure sufficiently good conditions of life for humans and animals in general. It requires us to stop caring about how we can further refine pleasures and joys, continuously improving and seeking to maximize wellbeing, and start caring about how to maintain a satisfactory quality of life for all.

Third, subjective wellbeing is known to be subject to various forms of *adaptation*. That is, over time, people tend to adjust to their situation to maintain a certain level of wellbeing. The initial boost in happiness that comes experiencing from experiencing an improvement in living conditions, like getting a pay raise or moving to a larger apartment, soon fades, as the improved condition becomes the new normal. Similarly, the effect of negative events, like income reduction or even illness, also wears off after some time. This phenomenon has been described by psychologists as "hedonic adaptation" and conceptualized as a form of habituation to sensory stimuli or desensitization (Brickman & Campbell, 1971), though seldom consistently so—it is also seen as a cognitive process that depends on a shift in awareness or expectations (DeHaan & Ryan, 2014; Sheldon & Lucas, 2014). A structurally similar phenomenon is that people are prone to form what has come to be known as *adaptive preferences*. They adjust their preferences to their available feasible options—"if you can't be with the one you love, love the one you're with," as Stephen Stills once sang—and come to terms with, and eventually to appreciate, their situation.

I think it is important to be aware of the breadth and diversity of different forms of wellbeing-relevant adaptation. It comprises much more than just changes in sensory receptivity or preference adjustment, for example, coping behavior, being aided by others and being shielded from harmful influences. And the categories of hedonic adaptation and adaptive preferences are themselves broad and undifferentiated and in need of further refinement, as different kinds of hedonic adaptation or adaptive preferences can have very different normative significance (Klausen et al., 2022).

I will elaborate this point further below, arguing that adaptation is not generally problematic and does not tell against subjective theories of wellbeing. But first we must consider why it is widely considered to be a problem, and particularly so in the context of climate change. The general worry is that adaptation changes—and, in cases of "positive," apparently wellbeing-restoring adaptation *lowers*—the standard according to which wellbeing is judged or perceived. This is perhaps most obvious in the case of adaptive preferences, where ambitious goals are substituted for more realistic ones. But something similar can be said about hedonic adaptation, as it leads to an appreciation of something that was initially experienced as unpleasant, for example, the soup that is served in the prison camp. According to those who find adaptation problematic, it is wrong to say of the person who has adapted that she has restored her quality of life to its former level. For she does not live as good a life as she did before. Rather she has *resigned*; she has come to accept a lower quality of life as the best she can currently get.

Hence, according to its critics, the propensity for adaptation means that subjective wellbeing can be very different from genuine quality of life, which must be conceived and measured according to more objective standards. People who don't know the quality of good carefully prepared food or a diverse and green natural environment may come to feel satisfied with fast food and digital entertainment. People who don't know or have no possibility of forming close personal relationships may come to think that they are doing quite well with feeling affection for commercial idols, pets, or digital companions. We can imagine future life in a world damaged by climate change to be similarly impoverished, with less biodiversity, less varied food and leisure opportunities, severe health problems, violent conflicts, etc., but in which people have adapted and so continue to say and think of themselves that they are doing quite well. They are like the inhabitants of Plato's cave who prefer mere shadows to the real things. Here is how Tim Mulgan presents the case against subjective wellbeing in relation to the prospect of climate change:

> What if future people prefer virtual reality to encounters with the real natural world? What if they would not thank us for preserving virgin rainforest rather than pouring all our energies into developing ever more intricate computer games? If distant future people have no experience of diverse ecosystems or a stable climate, if they cannot even imagine such things, how can they prefer them? But then how can we fault any present decision to destroy those good things?

> By contrast, proponents of an objective list theory of well-being can easily explain what is wrong with such wanton destruction. We must ensure that distant future people have access to desirable things (drinkable water, breathable air, stable climate, ecological diversity) whatever their actual preferences. If a connection to the natural world is intrinsically valuable, then human lives go better (and perhaps can only go well) when they instantiate that value.
>
> (Mulgan, 2015, 105)

According to Mulgan, future people may well, through adaptation, come to have lost their *epistemic access* to important goods, the values of which are still clear to us and should govern our efforts to conserve and protect existing features of the world.

Fourth, the above reasons for worrying about adaptation are often combined with a *moral* concern. This combination can be found in O'Neill, who warns against using subjective wellbeing to argue against the importance of growth:

> Problems of adaptation mean that arguments that appeal to subjective well-being fail to address problems of inequality both within and across generations. The absence of change in subjective well-being is indicative of the problem. A robust defence of the ... claim [that the ending of continuous economic growth is consistent with the maintenance or improvement of well-being] needs to appeal to objective state accounts of well-being.
>
> (O'Neill, 2017, 141)

> ... the appeal to subjective well-being is one that should be resisted as an argument for showing that well-being and growth can be de-coupled.
>
> (O'Neill, 2017, 146)

In arguing thus for the need to appeal to an objective account, O'Neill echoes concerns originally voiced by Amartya Sen in relation to global justice and human development. According to Sen, adaptation highlights the inadequacy of subjective wellbeing, because its application will be

> [u]nfair to those who are persistently deprived ... The deprived people tend to come to terms with their deprivation because of sheer necessity of survival ... and may even adjust their desires and expectations to what the unambitiously sees as feasible.
>
> (Sen, 2000, 63)

This objection from unfairness is often accompanied by an insistence that the appropriate response to grim living conditions is discontent and protest, not adaptation. If we accept that people may come to live satisfactory lives under objectively and morally deplorable conditions, then we lose the rationale for criticizing and fighting to improve them.

Fifth, expressing a combination of moral and prudential concerns, some find it objectionable that adaptive preferences have been formed *non-autonomously.* People come to value something they formerly didn't like so much, or to disvalue something they used to like. Jon Elster has famously criticized adaptive preferences for being a "sour grapes" response to unfortunate circumstances. He distinguishes between a change in preferences due to deliberate character planning and those that are formed without the subject's intention, and takes the latter to be problematic (Elster, 1983). This can be seen as a moral concern: people are *wronged* by being forced to develop preferences they would not themselves have chosen or wanted to cultivate if they had been given the choice. A similar concern may be implicit in some of the more general worries about adaptation to a future world harrowed by climate change: people will not know better; they will not have been able to choose anything else than the shallow gratifications with which they must now contend themselves, and they have developed a taste for desert landscapes simply through being exposed to them and nothing else.

3.3 Virtues of Subjective Wellbeing

I think the criticism of subjective wellbeing is misguided. I will argue that contrary to what the critics claim, subjective wellbeing is the most adequate notion of quality of life, *especially* for responding to climate change.

The critics are right that subjective wellbeing can be subtle, sensitive, and almost idiosyncratic. But this reflects the nature of human psychology and is something any viable approach to wellbeing must come to terms with. People get different kinds and levels of wellbeing out of not just trivially "instrumental" goods, like money or food, but also out of things like achievement, community membership, interpersonal relationships, outdoor and cultural life, and political participation (Lyubomirsky, 2001). People do not as a rule, however, respond very differently to, and they do tend to get significant wellbeing from most of the goods listed by the objective theories. Most people do take deep pleasure in close interpersonal relationships, though the preferred number, forms, and settings of such relationships differ. Many people from many cultures value achievement and accomplishment; maybe almost all people value it in *some* sense (as the Buddhist monk arguably values successfully completing a meditation session and experiencing a loss of self, and the Christian ascetic likewise values not taking any pride in her own actions, which are also kinds of accomplishment). There are robust patterns in the way people achieve wellbeing from specific activities and circumstances, and it would be misleading to say that it depends on subjective inclinations and preferences, even though the conditioning role of the latter should also be acknowledged.

When it comes to future life under conditions of climate change, however, we must be particularly cautious about judging it from the patterns we can now observe. For very much can have happened by then that will mediate or change people's responses to things that are currently conducive or detrimental to wellbeing. Maybe people will have found ways to appreciate new kinds of diversity, similar to the way we have already come to appreciate artificial (that is, cultural rather than natural) nuances and variation. Mulgan is right that drinkable water and breathable air are material preconditions for human life. But we should leave it open how these necessities will be made available to us, not ruling out in advance that we can be supplied with water and air through artificial means, as people already are in many locations and situations.[5]

But what about the gist of the objection, which is that because subjective wellbeing is subtle, diverse, and changing, it is too difficult to measure, predict, and base pro-environmental policies on? There is surely something to this. We should be mindful of the purpose for which we want to use the theories, and open to the possibility that different theories can serve different purposes. Objective list and needs theories provide good rough indicators of wellbeing. They may also help to correct some of the errors and inaccuracies that can befall studies based on self-reporting. If the typical conditions for love and friendship, peaceful coexistence, artistic expression, knowledge-seeking, etc., do not obtain, we should indeed be suspicious of a high level of self-reported happiness. On the other hand, notions of subjective wellbeing have been successfully operationalized, and the very large amount of empirical research made on this basis indicate that objective goods or objectively favorable situations may cause different levels of wellbeing due to various mediating factors. This indicates that it is indeed variable, but also shows that this makes it neither elusive nor resistant to practical interventions.

3.4 Growth Ideology and Individualism?

What about the claim that the concern for subjective wellbeing is part of the very growth ideology that has been, and still is, a central cause of the climate crisis? Should we not concentrate on ensuring that people's basic needs will be met? Even though there might be more to a perfect quality of life, the present situation does not call for actions aimed at further refining already subtle and resource-consuming joys of well-to-do people.

I will dismantle this criticism in three steps. I will argue that subjective wellbeing theory is not inextricably connected to material growth ideology, but function as an antidote to it. I will then try to show that "fundamental needs" is an ambiguous expression and that the concept of needs that is most often invoked has a much less clear meaning, and less obvious importance, than is usually assumed. Finally, I will argue that we do best without assuming that values and needs are *satiable*. Whether they are so is an empirical question. But it is most safe to assume that humans are capable of, and to some extent have a need for, improving their wellbeing regardless of how well they are doing already.

First, then, let us consider the ideology critique of subjective wellbeing. One of the things that seem to speak for it is that the concern for subjective wellbeing is central to a theory like utilitarianism, which has become a mainstay of classical liberal economics, which in turn has supported unlimited economic growth. It is also correct that current policy interest in subjective wellbeing is related to neoliberal concerns for maintaining an efficient workforce and keeping public spending within limits (Bache & Readon, 2016), and that the notion is usually applied to, and embodies a certain focus on, individuals (Elraz & McCabe, 2023; Krys et al., 2019).

However, these are superficial, contingent connections that should discredit the notion of subjective wellbeing as such. Ever since its emergence in the nineteenth century, utilitarianism has been associated with progressive social policies, and utilitarians have, in line with older hedonists like Epicurus, usually criticized the tendency to confuse material goods, like wealth and economic growth, with intrinsic goods. On the other hand, the concern for subjective wellbeing does not commit one to utilitarianism; it is integral to all major strands of ethical thinking. Notwithstanding its alleged neoliberal flair, contemporary wellbeing research and policy are dominated by a similar tendency. Subjective wellbeing is regularly invoked to support the view that wellbeing is not to any large degree dependent on material wealth or growth, but that trust, security, social relationships, experiences of autonomy and agency, etc., are more robust determinants of it. Hence the focus may initially be on the individual, but very soon becomes widened to include the societal conditions on which individual wellbeing depends. There is little basis for the accusation that subjective wellbeing makes the individual responsible for how well she is doing. It is true that it entails that securing certain living standard does not guarantee individual wellbeing. But it does not entail that it is up to the individuals themselves to muster the psychological and strategic resources needed for this.

Although the term "wellbeing" is, admittedly, most often applied to individuals, subjective wellbeing and the ability to foster and maintain it can just as well be ascribed to groups (Engelsen, 2022; Klausen, 2018b). Indeed, this may be an advantage of a suitably nuanced subjective wellbeing approach. For if we want to enable people to live both sustainably and well, it is probably advisable to focus on how groups of people—families and communities, for instance—can maintain and improve wellbeing by mutually adjusting their actions, behavior, and preferences.

It is a further relevant consideration that because subjective wellbeing can be supported by hedonic adaptation (see the next section), it is possible that it can be achieved with less material consumption than on an objectivist view, which stipulates more rigidly that a certain amount of goods or certain living conditions are necessary for living a satisfactory life. Yet while the critics of subjective wellbeing worry that it can license an acceptance of impoverished living conditions that ought not be considered good enough, their main worry seems to be, somewhat paradoxically, that subjective wellbeing goes hand in hand with an unsatiable demand for new pleasures and gratifications and so cannot be separated from the problematic growth ideology.

The question of whether fundamental prudential values are satiable or not is complicated, but important enough that we must consider it at length. There is, however, a simpler reply to the critics of subjective wellbeing that does not depend on any specific view on this contentious issue. For both psychological facts *and* morality make it extremely unlikely that subjective wellbeing should in practice support a growth ideology or justify giving special attention to costly or idiosyncratic pleasures. Subjective wellbeing research very strongly indicates that we get by far the most wellbeing out of improving people's general living conditions and tending to their basic needs. That it should be so is predicted by the economic principle of diminishing marginal utility: as consumption increases, the marginal utility derived from each additional item declines. That is, people get relatively less wellbeing out of the extra, for example, luxury goods, they consume, or additional possessions or income. If tending to special, artificially developed needs and preferences is costly, in economic or ecological terms, then we will be morally obliged not to do so.

This alone should suffice to reject the critics' claim that subjective wellbeing is wedded to growth ideology or supports an unfair or unsustainable distribution of resources. Regardless of whether the fundamental prudential values are satiable or not, psychological realism together with a concern for fairness and sustainability will ensure that our efforts will be directed at improving general living conditions rather than providing luxury or tending to subjective whims. This does not, however, render the question of satiability and possible limits or thresholds for concern irrelevant. Subjective wellbeing may not be as problematic as the critics claim. But it could still be that an objective wellbeing approach is best suited for use in the context of climate change, where the focus must, apparently, be on how to meet basic human needs.

3.5 Why Not a Needs Approach?

There is no denial that objective conceptions of quality of life, like the capability and the needs approach, do appear well suited for use in the context of climate change. For it does seem that the climate crisis requires us to concentrate on what is acceptable or sufficient, to keep the environmental impact of human life at a minimum. It also seems to be the most obvious way to justify and regulate the broader distribution of resources that is demanded by both the climate crisis and a more general concern for global fairness. The needs approach is considered inherently *frugal*, which is just what the climate and other environmental crises seem to call for. It is seen as providing a basic for distinguishing clearly between *necessities* and mere *luxuries* (Büchs & Koch, 2019). Moreover, the very notion of *sustainable development* was defined in terms of needs in the seminal Brundtland report (Brundtland, 1987), suggesting a close link between the overall goal of pro-environmental action and the notion of needs satisfaction (Guillén-Royo, 2018).

However, closer scrutiny of the idea of basic human needs and their satisfaction reveals that it has significant limitations. It is fundamentally vague and unclear and so insufficient for providing guidelines as to what to do in practice. On its most popular interpretations, it turns out to be not so obviously attractive. And when it is interpreted more sensibly, the concern for needs satisfaction and subjective wellbeing tend to converge. But this also means that there is room for reconciliation. I am open to eventually using both needs terminology and a suitably interpreted notion of needs to describe, assess, and improve situations related to climate change.

Now, what is a basic human need? It is sometimes used to refer to something that is physically necessary for human survival or the maintenance of basic health—air, food, water, etc. This is undeniably important to wellbeing since one must stay alive to live well. Yet satisfaction of needs in this sense (what we could call *subsistence requirements*) seems to be an instrumental rather than an intrinsic, prudential good. It is needed for one to be able to live and function at all and so enjoy a good life but does not seem to contribute to the goodness of life itself. That it is not intrinsically good can be seen from the fact that we can imagine situations in which it would better for a person to be deprived of one or more of these "necessities of life." This can be the case if her life is no longer worth living—that is, if she is no longer able to experience sufficient wellbeing. Passive euthanasia, that is, "allowing" a person to die because keeping her alive would only serve to prolong her suffering, is sometimes carried out by withdrawing tube feeding (and so no longer meeting a basic physical need). So although the notion of needs as subsistence requirements is as such sufficiently clear, it does not provide any guidance as to what is needed for life to be good enough.

In psychology, needs have been closely connected with drives, that is, biologically rooted inclinations, pressures, or urges for satisfaction; a famous example is Maslow's (1954) "hierarchy of needs." Yet there is clearly nothing intrinsically good in meeting needs in this sense. An organism's drive for procreation or offspring protection may not benefit itself. On the other hand, drives are usually thought of as stemming from a tension or imbalance, which will often be experienced as unpleasant or difficult to endure, but pleasant to have resolved. This reflects a typical ambiguity in talk about needs, which vacillates between referring to objective requirements for the existence and functioning of a human being and to the *experience* of cravings or desires (usually for ending or mitigating some state of lack or deprivation) and their satisfaction. The latter is more plausible candidate for an intrinsic good but makes the needs approach more akin to subjective wellbeing. It is also characteristic that the psychological understanding of needs as (or related to) drives posits very diverse ends or "satisfiers," for example, achievement, activity, affection, affiliation, curiosity, elimination, exploration, manipulation, maternity, pain avoidance, sex, and sleep (Encyclopedia Britannica, 2023). Some of these, like achievement, are standardly listed among putative objective goods; some (like sleep) are rather seen as physiological preconditions for a satisfactory life; and some, like the drive for manipulation, are considered neither intrinsically nor instrumentally good as such.

Sustainable human living, consumption, and development have been extensively analyzed and debated with use of the theory of fundamental human needs put forward by Manfred Max-Neef (1991). Max-Neef's Human Scale Development (HSD) framework comprises nine axiological categories: *subsistence*, *protection*, *affection*, *understanding*, *participation*, *idleness*, *creation*, *identity*, and *freedom*. Each of these categories can be expressed in the four existential categories: *being*, *having*, *doing*, and *interacting*. The resulting matrix provides a tool for categorizing different ways in which needs are pursued—or satisfied—in a society. This can then be subjected to a normative assessment: which "satisfiers"—that is, which practices, values, institutions, etc.—are the best overall, for example, by being ecologically and socially sustainable? Can we point to, or imagine *synergic*, satisfiers that contribute to meeting more than one human need and are not detrimental to any (Guillén-Royo, 2018).

It springs to mind that not all the HSD axiological categories refer to something that is normally considered an intrinsic good, although there is a good deal of overlap with typical objective list accounts of wellbeing. I have already criticized the idea that subsistence is an intrinsic good, and the intrinsic value of protection could also be questioned. It seems to depend on the context. Protection is more important the more vulnerable one is, and not so important if there are few significant threats in one's environment. This could be seen as an indication that needs are contingent and that their satisfaction can be rendered less important by changing features of a person's environment or psychology.

This apparently context-relative and changing importance of needs seems to clash with the basic theoretical assumptions behind the needs approach. For the needs in question are assumed to be *fundamental*. They should function as ultimate prudential reasons; that they ought to be fulfilled (*ceteris paribus*) is taken to be neither capable of nor requiring any further justification (Rauschmayer & Omann, 2015). Needs are also considered to be universal and persistent, changing only at the slow pace of human evolution (Guillén-Royo, 2018). They are, moreover, supposed to be *non-substitutionable*. Different dimensions of needs satisfaction cannot be traded off against each other (Büchs & Koch, 2019).

This underscores their presumed fundamentality or "essential" character: according to the needs approach, to live well enough, *all* one's basic needs must be fulfilled. Yet it does seem as if at least the relative importance of particular needs is subject to negotiation and could be lowered, perhaps considerably, by satisfying other needs.

It is a further theoretical implication of the needs approach that needs are supposed to be *satiable*: there is a threshold at which they can be satisfied and beyond which the use of additional resources does not add significantly to their satisfaction (Gough, 2017). As already noted, satiability is a notion that is especially pertinent to the issue of wellbeing and climate change, which often revolves around what would be good enough, what we could do with, what is superfluous to living a good life, etc. I have myself framed the central question of this book by asking how far it is possible to live a good and satisfying life under constraints caused by climate change. This indicates that there must be some degree of satisfaction that is *sufficient* and simply *satisfactory*; that we have an intuitive idea of what it takes to make a life go well enough; and that this is what we should concentrate on.

The paradigm instance of a satiable need is the desire or demand to end hunger (the "need for food"), which can be completely met: it requires one to eat a definite amount of food. The desire for status, on the other hand, is considered unsatiable, both because one could always in principle rise to new heights in power or comparative excellence and because status is relative and the reference group—the people one compares oneself to—is likely, in a society where status is valued, to also keep pushing ahead, thus perpetually raising the bar and creating and arms race and "treadmill" effect; in spite of all one's efforts, one's place in the status hierarchy remains roughly the same. The needs approach to quality of life uses something like hunger and its satisfaction as a model for understanding the requirements for a good life more generally.

Joseph Raz has elaborated the notion of satiable value by considering *happiness* as an example of it. Raz argues that *having a happy childhood* is a threshold concept, because it expresses a satiable value; if your childhood has been good enough, then you have had a happy childhood, period. It wouldn't have been any happier if you had had still more pleasant experiences (Raz, 1986, 242). Happiness is not *infinitely accumulative*, as hedonists will have it (Nielsen, 2019). There is such a thing as perfect happiness, but it is not a utopian state of having infinitely much of everything or excelling in all respects. It is the more mundane state of living up to what can be reasonably expected of a happy phase in one's life.

This is not wrong, but it is only part of the story. The word "happiness," like the word "wellbeing," is indeed commonly used to express what we might call a *threshold concept:* the concept of something for which there is a relatively clear threshold or standard of "enoughness," like satiation or satisfaction. We use it to describe a general, vaguely delineated positive long-term state of human beings. I suppose that this use of the concept is to a considerable extent determined by cultural stereotypes. As far as it can be stated more precisely, it may be roughly equivalent to having more than sufficient reason to be satisfied with one's life (Klausen, 2018b) or having a significant predominance of positive over negative experiences and emotions (Klausen, 2016). But we also have used the concept of happiness to refer to a quality that is, if not infinitely accumulative, then at least something that can in practice always be improved, and which is never so good that we should not, *ceteris paribus, care for it* and so be concerned with improving it.

When Raz speaks of having a happy childhood, he is using "happiness" in the threshold sense to describe a state roughly delineated by reasonable expectations and sufficient similarity to a stereotype. He is right that one could not have had "more" of a happy childhood in this sense. If you have it, then you have. But he is wrong in insinuating that a child enjoying a happy childhood could not have been any *happier.*[6] It is no excuse for not tending to or caring for your daughter's temporary depression that she otherwise has a happy childhood. So happiness in the sense which admits of no definite threshold is something we also should, and actually do, care for. And we mustn't bifurcate happiness into two distinctive concepts to accommodate both this and Raz's observation. For threshold concepts can be used to express values that are not completely satiable. It may suffice for the surface of my desk to count as being smooth that it has no profile irregularities beyond 100 micrometers. But this does not mean that it could not be still *smoother*, and this could become a practical concern if it were to be used for, for example, conducting mechanical experiments. Raz is wrong in claiming that adding pleasures to an already happy childhood would not make it happier. It does make sense to say that Rebecca had a happy childhood, but that Margot had a still happier childhood. This could be because Rebecca, though she lived well enough, overall, did go through some stressful periods or because she was deprived of certain experiences which, though not necessary for overall happiness, would have added to the quality of her life. It makes sense to care for happiness above the threshold for counting as happy overall also because the standards of needs satisfaction, and the stereotypes based on which we apply the concept of happiness, change over time. What made for a happy childhood fifty years ago may not make for one today.

The needs approach is supposed to provide a way around the changing expectations, preferences, and standards of comparison. But if the requirements *for actual need satisfaction* are highly variable, as seems to be the case, then the needs approach does not hold any significant advantage over subjective wellbeing. Both approaches must deal with the fact that needs can be satisfied better or worse, and that one can get a very different quality of life out of simply having one's need met.

Proponents of the needs approach often invoke a notion of human *flourishing* to set a standard that is both substantial and not context-dependent (Rauschmayer & Omann, 2015). But even if we agree that human needs do not vary much across times and places, or between individual people, it remains equally true that you can flourish in very different ways and to different degrees. The notion of flourishing is often taken to refer to something close to subsistence, like being able to go in living a decent human life and exercise one's basic human functions and capacities. But it is likewise often understood in a perfectionist sense, as fulfilling one's complete human potential (see, e.g., VanderWeele, 2017). The problem is not so much that needs satisfaction and flourishing are themselves context-sensitive, though this may also be the case (Fletcher, 2021, 119). It is that they allow of different instantiations which are far from equally good and bad.

The proponent of the needs approach can argue that we should look for those satisfiers that are compatible with, or supportive of, the satisfaction of other needs. But it is unlikely that this suffices to discriminate finely enough between different living arrangements. It may fare better in this respect if what is meant is that we should investigate *how strongly* the satisfaction of one need supports or detracts from the satisfaction of others. But such a more nuanced assessment is, again, likely to require something like the subjective wellbeing approach. How far satisfaction of the need for protection or participation will be impacted by, say, satisfying the need for affection or idleness must depend on the degree and kind of vulnerability of the target population or the political context. One could try to spell this out in still more detail, while still describing it in terms of objective states of affairs. But this will make the description no less complicated than one made in terms of consequences for subjective wellbeing.

The concern for sustainability and mutual compatibility of satisfiers may motivate us to try to restore a situation in which it takes less in terms of material resources to live a satisfactory life, like in older days, when less material consumption was the norm. It is of course agreed on all hands—also by the proponents of subjective wellbeing—that if the same amount of wellbeing can be achieved with fewer resources, we should prefer such a situation and try to bring it about. But what if the situation we are aiming to restore is rather one in which the norms and expectations were more modest, because less was available and fewer forms of enjoyment were know? Then it would seem that needs satisfaction is compatible with a lower level of actual wellbeing. Remember that the subjective wellbeing approach has been criticized for potentially sanctioning a life that is in effect less good, perhaps not even good enough, in terms of genuine wellbeing, because of adaptation effects. But now it seems that the needs approach could be accused of something quite similar: we could have two different life situations, which the theory judges to be equally good for the persons involved, but which differ considerably with respect to prudentially relevant features and so are in fact *not* equally good.

There are indications that some forms of frugality and simplicity, which in theory should be satisfactory, are not actually perceived to be so. Various modernist movements, like functionalism in architecture, have tried to do away with decoration, embellishment ornament, coziness, harmony, etc., but they have failed to gain wide popularity and been criticized for ignoring facts about human psychology. Though some of this criticism underestimates the capacity of humans for aesthetic innovation and appreciation of different artistic styles (Pinker, 2002), it cannot be denied that across cultures, people do strongly value forms of cultural expression that go way beyond the bare functional minimum and in many cases comprise artifacts or actions that have no practical purpose.

This also illustrates that it is not so easy to distinguish between necessities and luxuries as proponents of the needs approach would have it. Max-Neef's HDS theory recognizes *play* as a fundamental human need. But play is per definition doing something that is *not* about satisfying other needs or tending to necessities. According to one influential theory, play is both the origin of human culture more generally and characterized by *superabundance*, essentially going beyond physiological and psychological needs (Huizinga, 1949). The need for idleness seems to be in the same ballpark, maybe also the need for creation. There is much that indicates that we have a need [*sic*] for doing more than just satisfying our basic needs, a need for some kind of surplus or excess.

It is sometimes claimed that satiable values, including the value of happiness, must be *diminishing* (Nielsen, 2019). Raz defines this as the principle that partial satisfaction diminishes ones' claim to resources (Raz, 1986, 235). But this is a principle of distributive justice that must be distinguished from diminishing in the psychological and prudentially relevant sense, which is what matters to wellbeing. Maybe the fact that your hunger has already been partially satisfied diminishes your claim to food and obliges us to feed someone else, who are more needy, first. It is an altogether different question whether you could get a significant benefit out of more food. There will no doubts be limits to that as well. But different needs and values may have different satiation levels and profiles.

Moreover, diminishing is not necessarily *disappearance*. It may be so in the case of simpler needs like that for food or satisfaction of hunger. There is a satiation point beyond which it will not do you any good to eat more, period. But it is much less obvious that the same holds for a multidimensional, enormously complex, and multifariously instantiable value like that of happiness. True, you may get to a point where you have ample reason to be satisfied with your life, and so meet the criteria for happiness, period. But there will no doubt still be room for considerable improvement. You might cultivate new interests, acquire new tastes, deepen your (already sufficient) understanding or appreciation of things in the world, or further deepen and intensify your personal relationships. This may not make your life go *much* better. Diminishing marginal returns are surely to be expected. Still, this principle may kick in much later in the case of immaterial goods. Could one not benefit significantly from a little more love, or from getting a little more out of the love she already has in her life? The principle of diminishing marginal returns does not entail complete disappearance of returns, nor that the returns fade into complete insignificance. It entails that it will take more and more of a certain resource to get a perceptible, additional benefit out of it. That leaves it an open question whether it can still be worth trying to obtain more of the good in question.

Another popular way of explaining the alleged frugal nature of the needs approach is to say that it is focused on *eliminating serious harms* (Doyal & Gough, 1991). Subjective wellbeing is thought to justify a concern for making people's life still better, by satisfying their wishes or fostering new positive experiences. The needs approach, by contrast, is concerned with alleviating suffering, pain, unsafety, and whatever else may be a cause of illbeing. This tallies with the psychological understanding of needs, which explains them as grounding in tensions calling for a resolution (and usually accompanied by feelings of frustration that also call for being eliminated or alleviated). Again, hunger is a fitting example. Satisfying hunger is markedly different from, and more important than, fostering culinary experience. However, a proponent of the subjective wellbeing approach will agree that it is often most important to tend to needs in the negative sense, like getting rid of pain, suffering, and frustration. This priority is easily explained by the fact that doing so raises the level of wellbeing much more than tending to a wish for a special treat or particularly nice experience.

Proponents of the needs approach may insist that we have deeper reasons for giving priority to needs satisfaction in the negative sense, and that this priority must be absolute. It is widely believed that negative experiences are prudentially and morally more important than positive ones. Many tend to think that the relief of pain should take priority over the promotion of pleasure (Acton & Watkins, 2015). However, this asymmetry principle is question-begging, since it depends on how the prudential value is quantified in the first place. Surely an intensely negative experience trumps a mildly positive one. Yet to justify a genuine asymmetry between the values as such, a less intense or durable negative experience should be able to trump a more intense or positive experience due *solely* to its negativity. This is much less plausible. People gladly trade off significant pains against experiences of achievement or love or understanding. Especially if one accepts a pluralism about pleasures and pains, as qualitative hedonists do, it becomes impossible to quantify the amount of value that should be ascribed to an experience independently of assessing its specific quality.[7] My experiences of my children being born were relatively brief, complex—joyful, transformative, but also stressful—and very, very positive overall. They had a large value, despite their brevity and lack of "pleasure" in the narrow sense of the word.

However, the asymmetry question is too complicated to be resolved here. But it should be less controversial to note there is no clear difference, *in general*, between eliminating or reducing harms and fostering or raising wellbeing. In many cases, we can *describe* the same event in both ways, as an instance of meeting a need and eliminating harm *and* as an instance of contributing positively to a person's wellbeing. Consider, for example, sexual needs and sexual frustration. Surely this is, on the surface, a state of deprivation. It is about lacking or missing something, and what is prudentially required seems to be that the deprivation is eliminated, end of story. But you can hardly have your sexual needs met or your sexual frustration eliminated without experiencing something pleasant and thus positively good. In fact, merely having one's "lack of sex" eliminated seems like a paradigm example of *not* having one's sexual needs met! Consider also the need for interpersonal relationships. Does meeting it only require that a person's isolation or loneliness is eliminated? Does it not also, and more essentially, require that one comes to enjoy being related to and interacting with other people? Taking the good of people to consist simply in having their frustrations eliminated is a sign that one has an impoverished notion of wellbeing. This was forcefully noticed by Bernard Williams, who distinguished between desires that can be satisfied simply by extinguishing the desire itself, and what he called *categorical* desires, which can only be satisfied by achieving the desired object (Williams, 1973). Williams rightly took the possession and satisfaction of categorical desires to be essential to a life worth living. We want, or should want, to positively enjoy the things we desire.

This is not to deny that there is something to the emphasis, famously made by Stoics and Buddhists, on tranquility, imperturbability, and being free of pain, stress, and worries. But it is more appropriate to view this as highlighting the importance of a positive emotional state (and especially positive moods) than simply the absence of desires that might be frustrated. And it may be worth risking some pain and frustration to experience or achieve something that adds substantially to one life and wellbeing. Close interpersonal relationships, like love and friendship, may be the most obvious examples. There will no doubt be a point at which seeking further goods and pleasures will be counterproductive. Balance and equilibrium in one's psyche and existential economy are important for wellbeing. But such a state can be maintained in richer or poorer and more or less dynamic ways. The equilibrium can be expanded (see Section 3.8 below).

3.6 Genuinely Basic Needs?

I have been arguing that the needs approach should not be seen as an alternative to subjective wellbeing but must be supplemented or merged with an approach to wellbeing that takes the role experiences and attitudes seriously. Now there is in fact a needs theory which does exactly so. This further illustrates the confusion that surrounds the use of the concept, but also how natural it is to think of needs satisfaction as involving real *satisfaction*, that is, some kind of subjective gratification. The highly influential psychological self-determination theory (SDT), developed by Deci and Ryan, employs a notion of satisfaction of basic needs understood as "specific phenomenal satisfactions" (Ryan & Deci, 2017). Understood in this way, needs satisfaction is, of course, not independent of the subjects' experience and attitudes. The basic value phenomenon (which SDT also takes to be the fundamental causal factor in motivation and behavior) seems to be an *experiential quality.*

In other respects, the SDT notion of needs does seem to be "objective," for example, insofar as needs are conceived as being culturally invariant. It is also objective in the sense that the deprivation or satisfaction of needs is assumed to have measurable functional effects, effects that obtain regardless of a person's subjective goals and values (Ryan & Deci, 2017, 10). This sets it apart from a preference satisfaction view, but not from a more robust mental statist view like hedonism. SDT defines needs as "nutrients that are essential for growth, integrity and well-being" (loc. cit.). While the needs in question are clearly psychological, they are understood in analogy with the biological needs of plants for certain nutrients that enable them grow and functioning. SDT also employs the quasi-biological notion of "flourishing." This may seem to contradict my observation that SDT takes the basic value phenomenon to be an experiential quality. But the experiential quality is at least *a* basic value phenomenon. SDT may take overall psychological functioning to be still more basic but claims that this requires the person to have certain experiences with a particular phenomenology. The latter are necessary conditions of need satisfaction. That their role in SDT is even more central—its real cornerstone—is indicated by Ryan and Deci's claim that the proximal, phenomenal satisfactions "reflect, in the deepest sense, the essence of human thriving" (Ryan & Deci, 2017, 5).

The SDT list of basic needs is considerably briefer than the corresponding HSD list (or matrix). It comprises only three, very general, needs: a need for *autonomy*, understood as a sense that "one's behaviors are self-endorsed, or congruent with one's authentic interests and values"; a need for *competence*, understood as feeling of mastery, of being able to operate effectively within one's important life context; and a need for *relatedness*, the feeling of being socially connected (Ryan & Deci, 2017, 10–11).

The SDT version of the needs approach is much less vulnerable to the doubts and criticisms voiced above. It converges still more strongly on the subjective wellbeing approach. Ryan and Deci have rejected suggestions that they should add additional basic needs to their theory. The three basic needs are described so generally that each covers a very wide variety of domains and modes of satisfaction. They are hardly satiable, at least not so easily as physiological needs.

Can you get enough mastery or enough relatedness? You can get enough of particular social relations, for sure, but it is likely there will almost always be room for adding some relation that will increase your felt satisfaction, perhaps substituting a more satisfying relation for a less satisfying on. Of course, SDT also contains the familiar claim that all you basically need is what is necessary for you to flourish. But we have seen that flourishing does not provide any very determinate standard for setting a threshold; it may itself be a gradable property. This is implicitly conceded by Ryan and Deci, who speak of "differences in the degree of satisfaction of basic needs" (2017, 242), note that greater needs satisfaction is correlated with higher wellbeing (2017, 243), and cite evidence for the existence of an "upward spiral":

> [T]he more autonomous people's personal goals are, the more they will engage in higher quality goal-related behaviors, and the more the people will experience satisfaction of basic psychological needs and the positive affect associated with this need satisfaction, which in turn leaves them more able to pursue concordant goals. … [M]ore concordant, autonomous motivations lead to enhanced personality functioning and adjustment over time.
>
> (Ryan & Deci, 2017, 212)

This observation points to a further feature of the SDT needs approach that makes it congenial to my own take on subjective wellbeing: it is essentially *dynamic.* It acknowledges that people's perceptions of need satisfaction are derived from domain and situational experiences, but that these are in turn colored by general levels of satisfaction (Ryan & Deci, 2017, 246). These kinds of dynamic interaction effects are important factors in subjecting wellbeing. But recognizing their importance is a far cry from simply identifying basic goods or valuable types of need satisfaction. It underscores the point that whether, and to what extent, something is experienced as satisfactory (and thus whether, and to what extent, a psychological need is satisfied) is determined by a complex and dynamic interplay of factors, many of which are subjective.

SDT is wedded to a "eudaimonic" and officially opposed to a "hedonist" understanding of wellbeing (Ryan & Deci, 2017, Ch. 10). But this contrast is neither so significant nor so clear as Ryan and Deci seem to assume. It has mainly to do with an emphasis on certain *experiential domains.* SDT emphasizes experiences of *agency* and *social relations*, whereas hedonism, as Deci and Ryan understand it, conceives of wellbeing in terms of passive and solitary experiences.

Yet hedonism is not so restricted. It countenances all sorts of experiences and all potential objects of experience, caring only for how *good* or *positive* the experience in question is. There is reason to assume that the experiential domains emphasized by SDT are in fact the most important, because experiences of successful agency and successful social relationships are widely considered the most rewarding and gratifying.[8] As for the allegedly objective orientation of SDU, it is noteworthy that Ryan and Deci acknowledge that need satisfaction is relative and mediated:

> People may differ in terms of how subjectively salient [the] needs are or how centrally the needs are represented in their personal goals and lifestyles, and these individual differences might affect need satisfaction.
>
> (Ryan & Deci, 2017, 242)

People are different, and what satisfies their needs and to which extent it does so depend on their context, their social interactions, their personally traits and their subjective interpretations of their needs.

One issue remains, however, before we can conclude that a sensible needs approach converges on a sensible subjective wellbeing approach: the issue of *substitutionability*. The needs approach, even in the more subjective and flexible form it takes in SDT, does not allow that any basic need can remain unsatisfied or insufficiently satisfied if a person is to enjoy an acceptable quality of life. It does not allow that needs and their satisfaction be traded against each other (Ryan & Deci, 2017, 250).

Now I have already argued that a categorical prohibition against *any* form of trade-off is implausible. On the other hand, the case for substitutionability is gradually weakened when the field of putative basic needs is narrowed down to the most general and fundamental, as SDT does. The question is not so much whether we should reject complete substitutionability, but whether a subjective wellbeing approach commits us to accepting it in the first place. I do not think that it does. Only a primitive quantitative hedonism entails that it is just a matter getting enough of some uniform good. Qualitative hedonism may also, in principle, allow that having enough of a single kind of particularly valuable experience—say, deep scientific insight or aesthetic appreciation—can make further types of positive experience, and so other needs or life domains, unnecessary. Yet on any remotely psychologically realistic view, a person will in practice always benefit from experiencing satisfactions of different sorts, related to different life domains, and very likely not be able to achieve sufficient wellbeing without some variety in her activities and corresponding experiences. The completely solitary genius is unlikely to remain so deeply immersed in her scientific or artistic pursuit that she will not feel a need for other gratifications, for example, mastering everyday tasks or talking to others.

This concession to the needs approach should not be taken too far, however. It is not inconceivable, and not even unrealistic, that some people have less need for feeling socially connected, and that they can do well (and "flourish") by feeling autonomous and competent to a larger degree. Sheldon and Niemic (2006) have found evidence that *balance* in the three needs is important to wellness. This is highly plausible, and I will incorporate a concern for balance in my own theory of wellbeing and self-cultivation. It does not imply, however, that all three needs must be fulfilled to the same degree, but rather that their modes and degrees of satisfaction should be mutually adjusted. The highly intelligent, less socially needy person may have other balance points than a person who is simpleminded, extrovert, and to whom the need for social interaction is very salient.

What are the lessons we can take from this critical discussion of the needs approach? On the negative side, we can conclude that the needs approach is not as such a serious alternative to the concern for subjective wellbeing. It is not enough to enumerate basic human needs and see to it that they all be satisfied. Needs can be satisfied to different degrees and in different ways, and need satisfaction is mediated by other factors, some of which are likely to be subjective. It is an open, partly psychological, question to what extent basic prudential values are satiable. They are not likely to be so definitely satiable as proponents of the needs approach assume. Especially immaterial values may not have a clear satiation point in the same way as needs like hunger and thirst. There will surely be limits to how far we should care about meeting needs, also for immaterial values, above a certain threshold. But where this threshold is, and the relative importance of meeting allegedly "basic" needs versus caring for other, allegedly less basic, things, may shift with time and context, and can even vary between persons. On the other hand, the relationship between wellbeing and objective living conditions is in many respects quite stable, and we will seldom be wrong in trying to improve (or preserve) these conditions along roughly the lines suggested by the needs approach.

3.7 Adaptation for Good and for Bad

The criticism of subjective wellbeing based on concerns for *adaptation* is the most complex and challenging. Because of this, and because I will make positive use of the notion of adaptation in my theory of wellbeing for green transition, it deserves a detailed and thorough response.

First, it should be noted that although adaptation is indeed a pervasive phenomenon, it is by no means ubiquitous. In fact, it is hotly debated how far lasting changes in subjective wellbeing are possible, and some psychologists have come to dismiss the very idea of a fixed baseline or set point of life satisfaction (Headey et al., 2014). Recent state-of-the-art research indicates that adaptation is limited to a certain range of conditions. People may never adapt to more extreme circumstances (Lucas, 2007). Hence one way of responding to those who worry about adaptation is to take its possibility as an indication that a situation is not fundamentally bad or negative. If conditions get very bad in the future, adaptation will not be possible, and subjective wellbeing will remain a reliable indicator (or valid construct) of quality of life. Another correction of earlier assumptions regarding hedonic adaptation is less immediately relevant to the criticism of subjective wellbeing but will prove important to my positive theorizing: people may adapt to *discrete life events*, like winning in the lottery, which produces no lasting change in subjective wellbeing. But setting *life priorities* and making *behavioral choices* may significantly and permanently alter one's degree of life satisfaction (Headey et al., 2014). Whereas what happens to people may be subject to adaptation, changes in attitudes and values effected by themselves can lead to a lasting, adaptation-resistant change in wellbeing level.

Second, we should be wary not to confuse the prudential question of whether adaptation can be a means to genuine wellbeing with moral and political questions about fairness and desert. Critics like Sen and O'Neill might well be right that it would be problematic to distribute resources based on how much wellbeing people will get from them, because it could benefit those who have already received their share. But this is not the issue here. And we need not modify our notion of quality of life to address it. We can simply reject a welfarist or utilitarian approach to distributive justice and opt for a deontological or more strongly value-pluralist approach.

But third, it is well possible that we *should* allocate resources based on how much wellbeing people will get from them. In some cases, this seems like the natural thing to do. Human beings consume a lot of resources, and while it can certainly be argued that they consume too much and should leave more to other sentient beings, it is also usual to think that humans are at least entitled to what they need to live acceptable, and possibly full, human lives, because such lives are particularly valuable. They are able to get more out of their lives and make more of them than most animals are, and this gives them a special standing (Kagan, 2019). It is more controversial to argue that different people—members of the same species—have a different moral standing, based on their capacities for enjoying wellbeing. But this is also not what I am arguing. I am making the more practical point that it will be imprudent—and probably lack moral justification—to allocate resources without considering the extent to which they support the wellbeing of those who get them. Consider this example. Joan and Jill have a different sensitivity to noise. They both *hear* equally well—they are equally good at taking in acoustic information. But Joan is also good at not attending to, and not letting herself be irritated by, noise in her environment. She directs her attention to other, more important things. She is conscious of the noise, but she subdues it, making it remain phenomenologically in the background of her experience. Jill, in contrast, is easily aroused by noise, being unable to ignore it and getting irritated by it. Imagine that you know this and have two otherwise similar offices to allocate among them (all other things being equal). Would it be most right to decide by flipping coins, so both have an equal chance of getting the least noisy office? Or should you give the noisier office to Joan, knowing that she can cope much better with it? I think the latter is obviously the right thing to do. Distribution and allocation should be sensitive to the different propensities of individuals and groups for benefiting from favorable and coping with adverse circumstances. It might be subject to other constraints. But the ability to translate extrinsic goods and into intrinsic ones is one important parameter.

There is no reason to assume that expected adaptation will diminish our obligations toward future generations and justify using more resources now, at least not to a degree which should be considered problematic. Expected adaptation should play *some* role in our overall evaluation of the future situation of humankind. It can be part of a more comprehensive justification for not feeling obliged to completely sacrifice current wellbeing or preserve the environment in its current state at any price. But it is highly unlikely that expected adaptation will provide sufficient reason for thinking that we are not required to significantly reduce our greenhouse gas emissions, or that we should not be concerned about the quality of life of future generations and consider their wellbeing at serious risk. This is not only due to the limitations in the scope and power of adaptation, which I pointed out above. It is also because only some forms of adaptation can said to be genuinely supportive of subjective wellbeing. The proponent of subjective wellbeing can thus agree with the critics that some forms of expected "adaptation" should *not* be used to justify doing or caring less for future generations.

For there are, fourth, *many different forms and instances of adaptation*, with different normative significance.[9] We can distinguish the following forms, many of which are also practically relevant in a positive sense, because they can be used as part of strategies and tools for maintaining wellbeing in otherwise difficult conditions (and so belong the theory of wellbeing for green transition in a wider sense).

(1) Developing an immunity to stimuli (not *experiencing* a negative event as such, even though one is directly *exposed* to it).
(2) Remaining *unaffected* by experiences (experiencing negative events, but not *apprehending* them as such, or not letting them influence one's overall wellbeing).
(3) Ignoring or downplaying negative experiences ("shaking them off," even though the negative events are apprehended as such).
(4) Putting negative experiences in (a larger) *perspective* (contextualization and recontextualization).
(5) Preventing negative events from having a negative impact on one's wellbeing (e.g., by developing new preferences, adopting new values or cultivating new mental habits) (*coping i*).
(6) Integrating a potentially negative condition in one's self-conception (e.g., identifying with an impairment or seemingly inferior social position).
(7) Preventing negative events from taking place (by developing new behaviors or redesigning the environment) (*coping ii*).
(8) Being shielded (by others, or by mechanisms or structures in one's environment) from exposure to negative events (*assisted* adaptation).

These forms of adaptation differ with respect to how far they can be said to maintain wellbeing or—as the critics are prone to emphasize—make the person come to terms with a lower quality of life. Importantly, however, each form of adaptation can also be realized in ways that are better or worse. It seems that the most potentially problematic forms are (3), (4), and (5), (6). If parts of future life will feel really feel bad, it is at best mildly good news that we may be able to direct our attention away from it (as in (3)). If we manage to keep up a good mood because we assure ourselves that it could have been much worse, or that we could have been dead, as an instance of recontextualization (4), then this is also hardly a reason for real optimism. If we develop a preference for desert landscapes because these will be the best we know, this seems like a "sour grapes"-response. If we get so used to inhabiting the desert landscape that we cannot imagine another life or environment as being ours or good for us (as a way of integrating a putatively negative condition in our self-conception (6)), then this does seem like an example of having lowered our standard of judgment rather than maintained our former quality of life.

Notice, however, that all these judgments are also supported by the subjective wellbeing approach, especially if subjective wellbeing is conceived mostly in terms of experiences and emotions (that is, along the lines of hedonism and/or affective wellbeing). If something feels bad to a person, then it is bad to the person on such a view, regardless of what she comes to think or say about it. On a hybrid view that comprises both experiential and affective *and* cognitive wellbeing, a tendency to not think negatively about it might still be a *mitigating* factor, able to make it less bad. But this also seems quite intuitive.

Consider now that some of the forms of adaptation listed can hardly be considered inherently problematic and may even seem generally benign. Mitigating and adapting (in the more practical, material sense) to climate change are instances of coping as described in (7), and/or assisted adaptation as described in (8). Also, (1) and (2) do seem, almost by definition, to support wellbeing, as they prevent the person from being *harmed* by what she is being exposed to. This may be slightly more controversial, because a proponent of objective wellbeing might insist that, for example, being in an unfriendly environment or being forced to feed on something that is less natural or appropriate to human beings could still be said to constitute a harm. But I think that immunity and unaffectedness are generally considered beneficial, and not as a kind of self-deception or blindness to a predicament. The positive significance of such dispositions rather serves to make objectivism seem less credible. I do not wish for my children that they become so sensible that they will be affected by all sorts of objective deficiencies in their environment. I also do not wish that they become obtuse and, for example, not able to feel compassion or grief. But this can be seen as the price for feeling connected to and empathizing with others and enjoying close personal relationships.

Notice, further, that there may be other *instances* of the apparently more problematic forms of adaptation that are likewise benign (that is, which contribute to genuinely maintaining wellbeing). Remember the case of Joan and Jill mentioned above. Joan was gifted with or had developed (we cannot tell from the description alone) an ability to not attend to, and not being irritated by, noise. This is also a case of (3), but it does seem that Jill's ability positively supports her wellbeing. Consider also a case like that of John, who has been come to realize, at first despairingly, that due to injuries, he will never come to compete at the Olympic Games, despite his great sports talent. John gradually comes to think that this is not so important, after all. A life as an esteemed sports coach may be less stressful and just as good. This seems to be a case of genuine wellbeing *restoration*, but also of recontextualization, of coming to see things in a new—more appropriate!?—perspective (as in (4)). This indicates that there may be a more specific, positively normative notion of recontextualization, according to which it is *not* appropriate to take comfort from comparing one's situation with that of being dead or living a very bad life. That is, there are standards of appropriateness which recontextualization may or may not meet.[10]

Adapting by developing new *preferences* (5) has been much more extensively discussed, as it is roughly the same as forming the so-called *adaptive preferences* (see above). I have already argued that not all adaptive preferences have the same normative standing. Even Elster allows for preferences that are changed intentionally as part of deliberate character planning. This may not seem relevant to the case of living with climate change. The critics of subjective wellbeing have had only the prospect of unconsciously formed preferences in mind. But intentional preference formation should be considered, also as a practical adaptation strategy. When facing a crisis, one may prepare for it by working on adjusting one's preferences. Imagine your spouse has fallen seriously ill, and it seems that this will make you unable to live the outwardly active life that you have enjoyed so much. In such a case it seems prudentially rational to developing stronger preferences for things you can do at home, like reading or listening to music. Part of my suggestion for how we should respond to the climate crisis is that we should work deliberately and strategically on cultivating new preferences that support mitigation actions and can be satisfied under the expected new conditions of living. I say "we," not just "you" (or "I"), because I envisage a process in which we work collectively on such preference formation. This of course raises further concerns about autonomy, and ethical acceptability more generally, since it will comprise situations in which some people try to bring other people to form new preferences, without the latter's reflective endorsement. I will discuss these concerns in Chapter 6.

It is not, however, only deliberately planned preference formation that can be normatively innocuous or benign. Many of our preferences are formed by causal processes outside our control. We simply acquire them, for example, by being exposed to certain kinds of food or sport or music. Nevertheless, we do not criticize them, but take their satisfaction to be genuinely good for us (Bruckner, 2009). Hence, there must be something more distinctively problematic about an adaptive preference for it to be dismissed as not good for the person who has it.

Identifying with a negative condition (6) sounds more intrinsically problematic. Identifying with being poor or marginalized and treating this as the baseline for wellbeing are the paradigm example of self-abnegating adaptation. The critics of subjective wellbeing are afraid that climate change adaptation will be like that. But what should we make of the fact that many people with disabilities manage to maintain a high level of subjective wellbeing (Albrecht & Devlieger, 1999), exactly, it seems, by coming to "assess their life as it is, not as it could be" and taking their condition to be part of themselves (Schramme, 2013)? Wouldn't it be arrogant and supercilious to dismiss this as merely a convenient rationalization? Consider how disability activist John Hull describes his experience of going blind:

> As one goes deeper and deeper into blindness, the things which once were taken for granted, and which were then mourned over as they disappeared, and for which one tried in various ways to find compensation, in the end ceases to matter. Somehow, it no longer seems important what people look like. One cannot check at first hand the accuracy of these reports, they lose personal meaning and are relegated to the edge of awareness. They become irrelevant in the conduct of one's life. One begins to live by other interests, other values. One begins to take up residence in another world. I think that I may be beginning to understand what blindness is like.
>
> (Hull, 1990)

Notice that Hull speaks of "taking up residence in another world," indicating that it is incommensurable with the world of the seeing, constituting its own horizon of values and interests. The parallel to the worries about climate change adaptation should be obvious. Critics of subjective wellbeing worry that people in a future world marked by climate change will have lost important goods but no longer care about them. Hull confirms that one might cease to care but refuses to see this as a loss. To him, it is rather a transformative experience that opens new dimensions of value.

3.8 Richer and Poorer Lives

The above discussion indicates that adaptation can be powerful mechanism in maintaining genuine wellbeing. However, while this follows naturally from the subjective wellbeing approach, it does not sit easy with some of my earlier conclusions that were also supported by concerns for subjective wellbeing. I argued against the needs approach that it may be possible to meet allegedly basic needs without still not doing well enough. Now I argue, apparently, that one can do well enough with less.

Yet rather than revealing a contradiction in the subjective wellbeing approach or my use of it, this highlights a tension within the objectivist criticism of it. For we have seen that this criticism is driven by a fear that subjective wellbeing requires too much, that it is greedy and opulent, and that it allows for too little, that it is complacent and abnegating. The worry about maximizing, growth, and caring for subjective whims expresses the one and the worry about adaptation expresses the other of these two fears. There is no formal contradiction. It is possible that we do not need much in terms of subjective gratification but need more in terms of favorable objective living conditions.

We can maintain that it is possible to live just as well or even better than under present, "normal" conditions, with disabilities or other significant reductions in what we can or have (for example, less biological diversity), while acknowledging that we would be *still* better off, *all other things being equal*, by having retained the lost functions or goods. This allows for the possibility that some, perhaps many, people with disabilities are in fact better off than the majority, because they have cultivated experiences, interests, and passions not affected by their disability, to a larger extent than people without disabilities. Yet they would have been even better off if they had retained and made positive use of their functions while simultaneously cultivating the special experiences, interests, and passions. The ability to see or to enjoy real biological diversity constitutes *pro tanto* benefits. But their loss can be compensated by getting more out of what one is left with. It is telling that we seldom pity people for not making use of their whole repertoire of experiential modes or the types of action available to them. We assume that a woman of letters can do perfectly well even though she has never experienced the joy of competitive sports, or that a life devoted to nature and its preservation can be more than good enough, even if the person living it will never indulge in the pleasures of mathematics.

What has emerged is that the discussion about basic needs and adaptation is, at root, a discussion about the value of diversity, variation, richness, depth, and simplicity. These are the more generic qualities of life and its content that one might reasonably care about. Satisfaction of basic needs may or may not lead to a life that is sufficiently rich. Adaptation to adverse circumstances may or may not preserve sufficient richness and depth. Neither needs satisfaction nor adaptation is as such prudentially good or bad; it is the richness or poverty of life and experience, which needs satisfaction and adaptation give rise to that can be ascribed intrinsic value. And even this requires qualification. Not every kind of diversity is good; and what counts as diversity in the relevant sense can be difficult to determine. Rawls famously adopted an example from Stace (1944, 238) to illustrate the ramifications of viewing wellbeing as desire-satisfaction, asking us to imagine a brilliant Harvard mathematician, who develops a strong—and fully informed—desire to count the blades of grass on the lawns of Harvard (1971, 432). Most have followed Rawls in finding it obvious that a life of grass-counting will be insufficiently rich. But this is not because the mathematician will not have many different things in her life. She will encounter very many different blades of grades and many different numbers through her grass-counting pursuits. It is rather because this is the wrong kind of diversity and the wrong kind of simplicity. It does not provide sufficient richness or depth to life.

I think we must acknowledge that diversity and simplicity in the relevant sense must be considered irreducible qualitative, *sui generis* phenomena (as richness and depth obviously are). Because their value can be very difficult to calculate from independent information, we have in practice to rely on subjective judgments. If it feels sufficiently varied and deep, or someone says it feels so, then it probably is. This is a further reason for preferring the subjective wellbeing approach. The presence of certain objective goods can also be a reliable indicator of a satisfactory life. But much depends on whether these goods are experienced or used in ways that are sufficiently rich and rewarding, and to determine this requires more fine-grained and qualitative information.

I am not claiming that we should grant people absolute first-person authority about how rich or poor, deep or shallow their experiences or activities are. The objectivist approach remains right in its implication that people can be wrong about their own good (but, as I have emphasized repeatedly, plausible subjective wellbeing theories have the same implication). We have ways of critically assessing and modifying subjective reports. Intersubjective agreement and reproducibility are standard criteria that might help rule out idiosyncratic perceptions. What we should not do, however, is to measure the reliability of subjective reports against what "we" take to be "obviously" good or bad. The point of the subjective wellbeing approach is that we should be open to the possibility that activities other than those usually endorsed by the cultural elite can be rich and rewarding.

A good example is video games. They are often seen, especially by those who are not digital natives, as mere diversion or superficial fun, and, like other digital activities, contrasted unfavorably with life "in the real world." However, empirical wellbeing research indicates that video games can make users experience satisfaction of the needs for competence, relatedness, and autonomy (Przybylski et al., 2010; Ryan & Deci, 2017). Similarly, contemporary research on the bodily, affective, and agential significance of video game spaces indicate that video game spaces allow for experiences of agency, that they often facilitate new forms of sociality and affectivity, and that they allow for control when needed (Ekdahl & Osler, 2023; Ngyuen, 2020; Osler, 2021). It remains a question for further study and discussion whether video games and other virtual environments still do lack some important qualities. But it is becoming obvious that instinctive dismissals are based on ignorance about their potential for contributing to wellbeing. Few today would deny that listening to or performing jazz or sophisticated rock music can be deeply rewarding; yet not long ago, these genres were considered the epitome of shallow entertainment, and "sophisticated rock music" was considered a contradiction in terms.

But are there no objective constraints? Could every object, activity, or situation become a source of sufficient wellbeing? Although this may in principle be so, there are likely empirical constraints that make some things practically necessary to wellbeing and other things practically incompatible with it. This should serve to further meet the objectivists worry that subjective wellbeing can be achieved through all sorts of seemingly impoverished and outright appalling lives.

Most of the imaginary cases used by objectivist to discredit subjectivism can be dismissed as being *psychologically unrealistic.* They assume a simplemindedness that is fortunately not to be found in actual human beings. People cannot in fact get sufficient pleasure or satisfaction out of mere grass-counting or ice-cream eating or pot-smoking. Nomy Arpaly (2010) makes the point forcefully: "I will make the following claim about the normal adult human who receives full satisfaction in her life from keeping a goldfish: she does not exist." Jonathan Haidt seconds her, noting that lawn mower racing and flagpole sitting can be recognized as less rewarding activities even though one does not believe in standards of objective meaningfulness. Since they are not likely to support any deep or vital engagement, they will only be subjectively satisfying if they are embedded into a wider range of interests and activities (Haidt, 2010). Also recall the fact that there are limits to the situations people can adapt to (see Section 3.7 above). Regardless how far they might be able to modify their attitudes, some things will continue to harm them or not provide them with sufficient stimulation or inspiration.

As I have already noted in concession to a suitably modified needs approach, it is, moreover, likely that a person must experience some competence, relatedness, and autonomy to achieve a satisfactory degree of wellbeing. But I also noted that these are generic qualities that can be realized in many ways and supported or manifested by a welter of activities. In many cases where people become able to achieve genuine wellbeing from something that were not previously considered valuable, this is due to their finding ways to use it as a vehicle for experiencing competence, relatedness, or autonomy. This often leads to the development of new competences and new forms of relatedness and autonomy, mastery of ego-shooter games, sports strategy analysis, tidying up one' apartment, building online relationships, etc.

3.9 Getting the Most Out of It: Cultivation as the Key?

We do not have to worry that people will adapt to life in a future world damaged by climate change in ways that make them unreasonably or inauthentically happy. If the world becomes sufficiently damaged, life will be bad, also in terms of subjective wellbeing. So we have good reason to preserve much of what we value about the world today. Subjective wellbeing provides no license to relax our efforts to combat climate change.

On the other hand, we should welcome the fact that humans turn out to be amazingly adaptative and extremely creative when it comes to finding new ways of doing well, even under adverse circumstances. This provides justification for climate change mitigation measures. It gives us reasons for optimism about how life will be in a future with less consumption, restrictions on movement, and other changes that might otherwise make it seem unattractive. For we might realistically hope to be able to maintain most or all of what is substantially good in the activities we presently enjoy. We may be able to maintain the hunter-gatherer-like excitement associated with shopping, while avoiding overconsumption and supporting more sustainable forms of clothing manufacturing. We may be able to continue expanding our horizon and exploring new places and cultural and natural gems without embarking on long flights. We may be able to enjoy culinary and gustatory diversity without consuming large amounts of meat, and so on. Some of this will be due to innovations on the *supply* side: new sustainable manufacturing methods, culinary innovations, the resurgence of night trains, etc. But the cultivation of our tastes and interests will be at least just as important. Indeed, we may not only maintain but come to genuinely improve our wellbeing. The restrictions on resource consumption, outward expansion and quantitative growth and variation may prompt us to deepen and refine our more fundamental sources of wellbeing, freeing us of the pressure to pursue activities that are not intrinsically satisfying.

The reasons for this optimism are not only theoretical. We have also seen that people with acquired disabilities tend to describe their lives in very similar terms: seemingly deprived, but in fact just as rich and rewarding, if not more, than the life they had before sustaining their injury. That we do not engage in current work and consumption activities out of immediate subjective desire can be seen from the fact that an enormous amount of money and efforts are spent on persuading us to do so (Soper, 2020). We are not faced with the task of combating natural, uncultivated inclinations, but rather of recultivating interests. And the fact that advertisement works may be regrettable, considering what it has been used for so far, but is proof of the malleability of human tastes and preferences.

The recent COVID-19 pandemic provides further evidence. The restrictions on daily life and movement were not generally perceived as fundamental obstacles to wellbeing. People employed coping strategies and were in many cases able to cultivate interests like reading or gardening (Lindgren et al., 2023; Mau et al., 2022). Domestic tourism boomed. People reported newfound joy in exploring their immediate environment or using proximate greenspaces (Lu et al., 2021) and developed preferences for more responsible travel seem to have (Orden-Mejía et al., 2022). The evidence from the pandemic is, admittedly, far from unambiguous (Hansel et al., 2022). During the second lockdown, people slipped back into old behaviors (Fancourt et al., 2022). Different groups of people were impacted and reacted differently, highlighting the importance of personal and social conditions for successful adaptation. The post-pandemic travel boom could be taken to indicate that apparent adaptations to more restricted ways of living were only temporary suppression of needs that had to be compensated in the longer term. Yet life during the pandemic did provide evidence that it is *possible* for people to improve their wellbeing despite, and partly even due to, external restrictions, and that cultivation and recultivation of tastes and interests can be a powerful tool for coping with new circumstances.

The significance of cultivating activities and experiences has received surprisingly scant attention in philosophy and wellbeing theory. An interesting exception can be found in the early work of Kierkegaard. In a short essay entitled "Crop Rotation," one of several pseudonymous texts that make up the first volume of his complex philosophical novel *Either/Or*, Kierkegaard develops the idea that the key to maintaining wellbeing lies not in outward expansion of one's activities or possession. Using the metaphor of extensive versus intensive agriculture, Kierkegaard argues that real satisfaction comes from cultivating and refining the more immediate affordances of life. Kierkegaard also explicitly claims that self-limitation can be a source of wellbeing competence:

> The method I propose consists not in changing the soil but, as in the real rotation of crops, in changing the method of cultivation and type of grain. Here, straightaway, we have the principle of limitation, which is the only saving one in the world. The more you limit yourself, the more resourceful you become.
> (Kierkegaard, 1992, 233; 2013, 281)

Kierkegaard lets his pseudonymous character, "the aesthete," express this idea, and he probably does not endorse it without reservation. He treats it satirically, for example, when he lets the aesthete describe enthusiastically how he can take pleasure from inciting a boring speaker to give a lecture, only to observe the sweat on his brow and his nose (2013, 288). This can be seen as indirect confirmation of the point that not everything can become a source of sufficient wellbeing, regardless of how creatively one approaches it. It should also be noted that Kierkegaard considers aesthetic satisfaction insufficient for living a generally satisfactory life. He does, however, see it as necessary, and as closely intertwined with the further necessary elements (as laid out in the second, "ethical" part of *Either/Or*). And he does keep to the general idea that limitation can support rather than hamper wellbeing, and that intensive cultivation and refinement of seemingly humdrum or familiar experiences can be more rewarding than chasing new and spectacular ones. This is an important lesson for green transition. We can learn to get much out of little. And this must not only be second best, a sort of compensation or consolation. The limitations imposed upon us by climate change may provide a golden opportunity to enrich our lives, enhancing their depth and internal variation.

3.10 Putting It Together: A Theory of Wellbeing for Green Transition

My assessment of different approaches to wellbeing may not have been sufficient to pick one very specific theory, still less to settle all the fundamental philosophical issues. Fortunately, that is also not necessary for my purpose. There is a virtue in being ecumenical, and we should now be able to determine what an appropriate theory of wellbeing will look like. We should be further able to specify what factors and elements will, according to such a theory, very likely be conducive or detrimental to wellbeing, thus paving the way for its application to pro-environmental behavior change. The theory should be focused on the *process* of trying or coming to lead a good life, rather than on characterizing a good human life from a detached standpoint. It should be a largely, though not necessarily completely, *mental state* theory, which takes wellbeing to consist to a large degree in having a favorable balance of positive over negative experiences and other mental states with a distinctive qualitative feel, not least moods and emotions.

For practical reasons, however, it might be preferable to adopt a moderately *hybrid* theory, which acknowledges the impact on wellbeing of satisfaction of likings, desires, and preferences, maybe even of some objective features of a human life, for example, the life context in which certain experiences are had.[11] For the satisfaction of likings, desires, and preferences is no doubt instrumental in maintaining a positive mental state. They are thus especially relevant for the *process* of living well. A central, if not the most central, task in this process will be to bring one's experiences, emotional and behavioral dispositions, *and* one's likings, desires, and preferences to accord with one's situation, as far as possible (Tiberius, 2018).

Although preferences may not be fundamentally constitutive of wellbeing, they are fundamental *tools* for achieving it. It is hardly imaginable that one could experience an almost complete frustration of one's wishes and still maintain a favorable balance of positive over negative moods and experiences. Weimann and colleagues have argued against adopting a one-sidedly affective conception of wellbeing, noting that conditions like unemployment take a significant toll on people's cognitive wellbeing and remarking that it would be inappropriate not to care for this (Weimann et al., 2015). I think they are right. But this is not because we should care much for the tendency to make a negative judgment of one's life if everything else is really going quite well. It is rather because a persistent and strongly negative judgment of one's life situation will inevitably make its mark on one's emotional state. And it is likely that the kind of negative emotions and, especially, moods, which are caused by and/or causes negative life satisfaction judgments, are not easily elicited by standard emotion sampling methodology. Life satisfaction may be understood not as a distinctively cognitive state, but as a deep, mood-like feeling of satisfaction or dissatisfaction with one's life. As such, it may still be partly cognitive, comprising a *perception* of how things are going in one's life. It is also worth noting that cognitive attitudes arguably have a characteristic phenomenology (Klausen, 2008; 2013). In the case of cognitive attitudes toward one's life and how well it is going, their phenomenology may resemble that of affective states and at least also be prudentially relevant.

This entanglement of cognitive and affective wellbeing is a good example of how wellbeing is the product of dynamic interactions between various states and factors. This supports a *holistic* view of wellbeing. We cannot assess the wellbeing of a person just by enumerating the different elements that pertain to her wellbeing, registering their individual contribution and then adding it all up. For the specific combination of elements also matters. We need not commit ourselves to the more controversial philosophical view that the prudential value of an element in wellbeing is holistic in the strong sense that its value is *non-separable*—that the elements cannot be evaluated independently of each other (Bader, 2016). It is sufficient to notice that the elements will influence each other causally, and very closely so. The value of a person's experiences will depend on her likings and preferences, which will in turn be conditioned and shaped by her experiences. Her emotional states will likewise both color her experiences and be impacted by them, as fearful experiences may cause stress and anxiety, and depression will make otherwise pleasant experiences—say, of good food or natural sceneries—much less pleasurable.

Preference formation is a tool for achieving wellbeing, but it is only one among many other processes and strategies that lend themselves to this task. Cultivating emotional and behavioral dispositions is another, often still more efficient tool. Seeking positive experiences may be the most familiar and obvious strategy. But because not all pleasurable experiences are deep and lasting, and because one may not be sufficiently mindful of which experiences best matches one's likings and dispositions, it is also far from a surefire strategy. It may be overrated, not because the value of genuinely pleasurable experiences is so, but because of the difficulties in having and cultivating them. Acting in new ways and shaping one's environment are a further type of wellbeing-supportive process. One can seek new social relations, extend and deepen existing ones, and avoid places and situations that are less good for one. All these mechanisms and strategies are relevant for how to lead one's life, but just as relevant for how to assist other people in achieving wellbeing. For they are all forms of benign adaptation that can be nurtured or prompted from without. Instead of letting it up to a person to forge interpersonal relationships, one can approach the person and introduce her to a certain group or community or see to it that others do so. One can contribute to making situation more supportive of a person's wellbeing, for example, by not offering alcohol or even having it in the house if a person is prone to drinking too much, or by painting the residence home ward in aesthetically pleasing colors.

The dynamic, flexible, and holistic character of wellbeing makes it *multiply realizable.* Two persons can do equally well but differ with respect to their degree of preference satisfaction or emotional state. Wellbeing has very different manifestations, and it *can* manifest itself in still many more ways. While these can be expected to exhibit some similarities, as they will surely involve some form of felt competence, autonomy, and relatedness, we should not imagine that we will be able to determine in advance how good or bad life under certain conditions will be. We should acknowledge that humans are highly creative when it comes to finding ways of achieving wellbeing, and that one can get very different kinds and levels of wellbeing out of the same objective situations and resources.

The elements in wellbeing are largely (but probably not completely) *substitutional.* Enough of one can compensate for the—almost complete—lack of the other. But one can achieve still more wellbeing by *adding* further elements and dimensions. A person who only enjoys solitary contemplation and aesthetic experience, but to such an extent that we should say of her that she is quite happy, would do still better if she could also come to enjoy conversations, friendship, craft, sport, political participation, or sex. This is not to say that she must strive for expanding the range of elements and dimensions of her wellbeing or that others should necessarily encourage her to do so. There is significant risk that her emotional dispositions are not well suited for other activities, or that her values make her averse to them, so that they will not contribute significantly to her wellbeing. There is also a risk that she will get less value out of her contemplative activities, no longer being able to devote sufficient time or concentration to them. The point is only that if she can expand her range of wellbeing-conducive activities and experiences, this will, *ceteris paribus*, enlarge her wellbeing. Moreover, while there may be significant psychological obstacles to successfully pursuing a policy of fast and radical expansion, there is almost always room for some enlargement. The solitary scholar should be able to take some pleasure from conversation or friendship, without losing too much. In practice, such expansion is often the most efficient way to improve wellbeing. For people will already have optimized their strategies for achieving wellbeing within their preferred fields of activity and experience and so face diminishing marginal utility.

This may sound problematic in the context of responding to climate change. It seems to contradict the optimistic point made above, viz. that we may be able to get more out of less and that limitations can be a blessing. Does it not mean that improving wellbeing requires growth after all? Not in any problematic sense. The point is not that the most efficient way to improve wellbeing is to consume more material resources. This way lies strongly diminished utility. The point is that one gets, *ceteris paribus*, more out of opening new fundamental life and experience domains. However, most people will already to some extent be engaged in many or most such domains, enjoying both interpersonal relations and different kinds of sensory pleasure, as well as exercising competence and autonomy in a range of different fields. For them, further cultivating some of these activities will indeed be most rewarding.

We should also keep in mind the importance of maintaining a balance between one's dispositions, inclinations, likings, preferences, and experiences. Keeping them aligned and mutually adjusted is even more important than enlarging the range of wellbeing-conducive activities. The subjective wellbeing approach thus converges on the needs approach in advocating a (relatively) steady, balanced form of life rather than one feverish expansion, though it differs from it in acknowledging the value and possibility of further refinement and enrichment within the bounds of sustainability.

It might be objected that I have not really presented a fleshed-out conception of wellbeing. I have not identified many specific constituents or central causes of wellbeing. I have merely outlined general structures and principles and described some highly generic factors. But this reflects the nature of wellbeing. It just is flexible and multiple realizable. Pinning it further down would repeat the mistake of some objectivist approaches of treating what is only one possible form of satisfactory life as if it were the only one. This would be particularly problematic in the context of climate change, where we have to consider life under future conditions that are difficult to predict. Besides, though I have not been very specific about the content of a satisfactory life, I have described several concrete tools for maintaining and strengthening wellbeing, in the form of different, individual and collective, adaptation and cultivation mechanisms and strategies. This is in keeping with the idea that a theory of wellbeing for green transition should be process-oriented, focusing on ways to achieve wellbeing rather than fixed ideals, contents, and conditions.

The fact that wellbeing is multiple realizable has important consequences for understanding and assessing the *changes* related to the climate crisis and green transition. It has already emerged (in Chapter 2) that changes may appear less impressive if the *forms* and *structures* of activities are preserved. This holds for example for food culture and transportation. It does not matter so much whether what you eat is meat, or whether your vehicle is driven by a combustion engine, as long as you can enjoy a varied and aesthetically pleasing meal or are capable of private transportation with a certain speed. This is now further supported by our inquiries into the nature of wellbeing: it is *how* you do it, or *what* you do *with it*, rather than *what* you do or what you do *it with*, that matters.

Here it might seem that I have come close to abandoning the very idea of *substantial* subjective wellbeing for a still more formal approach. Some have argued that because we cannot know in advance which goods will be appreciated by future people under conditions of climate change, we should concentrate on ensuring possibilities for action (Arler, 2014). My emphasis on what people can do also sounds like a concession to the capabilities approach of Sen, which is similarly motivated by a recognition that not only the presence or quantity of goods, but also the personal characteristics that govern the conversion of goods into a person's ability to promote her ends should be taken into account (Sen, 2000, 74).

But although an appropriate theory of wellbeing for climate change should indeed be flexible, it should still also be substantial. The theory I am proposing is process-oriented, but not procedural or formal. It does not reduce wellbeing to structures or functions, to the outcome of processes of deliberation or to abilities or opportunities for action. For just as Sen legitimately worries that basic goods might not be taken up or used in ways that benefit the persons in question, we should also worry that opportunities for action are not cashed out into genuine wellbeing, or that functions are maintained and exercised, but in ways that feel less satisfactory. This is not just a theoretical concern. It is an actual risk, as we already know from historical experience that seemingly well-designed environments or rational ways of organizing social life can feel subjectively unsatisfactory.

Moreover, flexibility is required because we can know relatively little about the future forms and contents of a good life. Yet our epistemic limitations should not be exaggerated. We can be sure that if people living under conditions of climate change will be stressed or depressed, not feeling attuned to their environment or able to control and manage their life, or not having many pleasant and varied experiences, then they will not be doing well. So we should do what we can to prevent such a situation. By contrast, possibilities for action are not much worth in themselves, and ensuring a large option or action space will be a very costly strategy. Not only will it be risky and wasteful, as options may prove irrelevant; extensive option freedom is arguably detrimental, rather than conducive to wellbeing (Haybron, 2008; Schwartz, 2004).

A further reason for not buying into a wholesale structuralism or functionalism is that the material aspects of practices and experiences also matter. The feel and sight of wood, the feel and smell of grass cannot easily be substituted. Specific kinds of gustatory or tactile experiences may be if not necessary, then at least important to living a fully satisfactory life. Notice, however, that the latter are still described in general terms, referring to types of experiences (that is, sensory modalities), leaving it open from what and how more precisely the experiences are to be had.

The concern for subjective wellbeing, and especially for people's emotional state and the quality and variety of their experiences, does strike an appropriate balance between form and substance, and between epistemic humility and confidence. And so far, it does seem that lifestyle changes in the wake of climate change need not pose any threat to our future wellbeing and may even be conducive to it.

Notes

1 Alexandrova (2017) argues that if wellbeing theory is to be practically relevant, it should not discriminate too strongly between constitutive and causal factors.
2 The capabilities approach is sometimes presented as a conceptualization of wellbeing (Robeyns & Byskov, 2020), perhaps more fittingly as an "alternative operationalization" of the concept (van der Deijl, 2020). Likewise, proponents of theories of the good life might also insist that they are putting forth a view of wellbeing, albeit one that includes a wider set of factors and qualities. However, the concept of needs satisfaction is more obviously different from that of wellbeing.
3 Haybron (2008) also adds an authenticity requirement in order to rule out that delusive or deceptive experiences can add to wellbeing. This is a further way of modifying hedonism which may make it more acceptable to many.
4 However, Lin (2019) has argued that if we focus on not just the satisfaction of a single, salient preference, but on the satisfaction of *all* the preferences a person has (e.g., not just for consuming drugs, but also for maintaining control of life an achieving a number of life goals), then theories of wellbeing as the satisfaction of actual preferences can deliver much more intuitively correct verdicts.
5 I write this while living in Beijing, where breathable air and drinking water are far from naturally abundant. It is surely not a wellbeing paradise, but clearly more livable than many locations that may come closer to Mulgan's ideal.
6 In fairness to Raz, it should be said that he is of course aware that one could have had more and more intense pleasures, but he does not take this to have any bearing on happiness as such.
7 Shriver (2014, 152–62). https://doi.org/10.1017/S0963180113000686 has argued that pleasure and pain are radically different states that also play different motivational roles, and used this to support the axiological asymmetry thesis. Yet the neuroscientific evidence he uses to support this view is largely irrelevant to the issue if one takes "pain" and "pleasure" to be umbrella terms covering a wide variety of negative and positive experiences and states.
8 Deci and Ryan also give some further normative reasons for rejecting hedonism as theory of wellbeing, for example, that being well also comprises dispositions (e.g., grounded in bad health) that are not manifested in a person's experiences, or

that feeling bad can be good for a person, for example, if the person is grieving the loss of a loved one.

9 The following paragraphs draw on Klausen et al. (2021).

10 Calhoun (2018) argues similarly that contentment is a virtue, and that one can cultivate more appropriate "expectation frames." She also acknowledges, however, that one can expect too little, or otherwise use "normatively distorted expectation frames" (164).

11 For example, pains may be less bad if they are endured as part of an autonomously chosen endeavor and not simply inflicted on one.

4

Living with Change

There is much that indicates that we should be able to achieve sufficient wellbeing under altered living conditions. However, an important question is still left open: What about the impact of *change* itself? There are three kinds of question to be considered.

First, we can ask whether the changes will affect our identity or life trajectory in a way that could be considered harmful and so should be taken into account. We have touched upon this at different places. Some worry, and some may actually feel, that they are about to lose important parts of their identity if they are no longer allowed to eat large pieces of red meat or drive their own car. Others fear that we may come to form new preferences in ways that are not sufficiently autonomous, either through climate mitigation measures or through the kind of hedonic adaptation that may accompany climate change itself.

Second, there are more psychological questions of how well we are able to cope with change. Is change itself perceived negatively or positively—and if so, when and how? Do we have a *need* for change, for example, in the form of growth, expansion, mobility, or diversion, which may be difficult to meet under more austere living conditions? Will climate change and the changes in behavior needed to keep it at bay make life more monotonous? The complexity of the latter set of questions reflects the general fact that change seems to be both a challenge, if not a threat, to human wellbeing, *and* a blessing, if not a necessity for doing well. Changes seem scary and are often hard to go through; but life without change seems boring and poor.

Third, there is the practical question of *how* to cope with or positively exploit changes. Which are the most appropriate ways to respond to changes that befall us or opportunities for personal transformation? How can we obtain sufficient variation and diversion while also maintaining balance and stability in life?

4.1 Remaining, Losing, and Becoming Yourself

Being and remaining oneself is widely considered to have fundamental value. It is valued so strongly that many seem to assume that it is even more important—that it has more prudential value—than achieving happiness. This is standardly assumed by existentialist philosophers (for example, de Beauvoir, 2015), but held much more widely, as can be seen for instance by its popularity as a theme in literary fiction.[1] Preservation of one's identity and continuity in one's pursuits is taken to be normatively significant to such an extent that it is seen as setting limits to how radically a person ought to change. Choices and changes are required to be "authentic." They ought to reflect one's basic values and beliefs (Varga, 2011), whereas actions that do not seem to reflect a person's self and character traits are viewed negatively (Paul, 2014). Calhoun observes that we value not just realizing the ends of practical activities but also *being a certain kind of person* in the process (2018, 75).

Some have argued that even if you were able to change your values and desires thoroughly and so be able to enjoy more and more varied goods, this would not really be good for *you*, because the good things would be enjoyed by someone very different from the person you are now (Kagan, 2012). Belshaw (2014, 2016) has made a similar point in response to Nozick's experience machine, both worrying that enjoying all the new superior experiences provided by the machine will come at the price of effectively *killing* the person that decided to plug on to the machine in the first place, and so cannot be said to be good for *her*. A similar line of argument has been presented by Robert Adams, who uses the case of Helen Keller to illustrate his point. Keller could not, Adams argues, reasonably have wished for never having been blind or deaf; for her never having been blind of deaf would have been like her never having existed (Adams, 1979).

Meghan Sullivan has named this kind of view the *conservation constraint* on preferences. It says that an agent can be rational in preferring something that actually happened to more valuable, but merely possible events (Sullivan, 2018). The preference for one's actual past existence may be thought to ground a constraint on preferences for one's future existence: just like wishing for having been very different can seem almost equal to wishing for having not existed, wishing for *becoming* very different can seem almost equal to wishing for one to no longer exist, and so almost suicidal. More pertinent to the case of green transition, *expecting* to become very different may seem close to expecting that one will cease to exist.

I will not assume that the conservation constraint is a fundamental truth about what is good for humans, or that authenticity has fundamental, intrinsic value. It might be said that our preferences for remaining ourselves reflect a *status quo* bias. Some people may be better off not clinging to their old personality but daring to let go of it. If my future life is going to be much better, and I (as a subject of experience) will still be there to enjoy it, why should I care about the degree of continuity with my former life and self? On the other hand, the concerns for conservation, continuity, and authenticity are very widely shared, and seem to be deeply ingrained in our thinking and emotional responses. Even when it comes to artifacts, we do tend to consider them worthy of preservation often simply in virtue of having been there and played a role, as can be seen from the large value attributed to cultural heritage conservation.

I am here considering the potential value of personal continuity, using our affectionate, preservative attitude toward cultural artifacts only as an analogy. But it should be noted that one could also use the value attributed to cultural heritage to argue directly against radical climate change measures. Arguments along these lines have surfaced in the debate, most clearly when it comes to how sustainability concerns may destroy local agricultural traditions. Defenders of the existing agricultural production have claimed that it should be recognized as an important "vehicle for culture and identity" on a national level (Christensen, 2020). On a smaller and seemingly less serious scale, some people have shown concern for fossil car culture and combustion engines, which they see as important cultural artifacts. This kind of thinking may also influence more general, including national, policies. One should not underestimate the role of nostalgic emotions in shaping official politics. The persistent defense of the fossil car industry in German politics has been, rightly I think, suspected of representing a peculiar German "vehicular nostalgia" (Kirkland, 2020).

Apart from seeing authenticity as intrinsically valuable, there may be instrumental reasons for favoring conservation and continuity. Stable personal identities and stable preferences make people predictable and dependable. They can also be good for people themselves, as they enable them to make genuine commitments and bring their projects to a conclusion. Changing interests and orientations may in some cases help people to enjoy new and better things. But in practice, they often seem to indicate a lack of perseverance or concentration and prevent people from getting enough out of the shifting things to which they turn their attention. There is also a risk that the objects of their new interests are not really "for them," that is, do not match their nature, their fundamental dispositions and inclinations, which may not have changed accordingly. This might suffice to uphold a concern for authenticity. But it should not be given too much weight. A case could still be made that it reflects a conventional prejudice or irrational bias. Authenticity may be good for some people; maybe it is possible to live well in less stable or continuous ways (Calhoun, 2018; Strawson, 2004).

Moreover, even if the general concern for authenticity deserves to be taken seriously, its relevance to green transition could be disputed. Although some might fear that their future life will be quite different, the changes induced by climate change and our responses to it are not likely to alter the values and preferences that are constitutive of us as persons (see Chapter 2). We do not literally face our own annihilation. A more likely worry is that we will, basically, remain ourselves, but undergo transformations so significant that we may feel prospectively alienated from ourselves. This is not just a luxury or academic problem. The so-called *future anxiety* is a widespread phenomenon. It has been characterized as a way of thinking about the future dominated by negative cognitive and emotional processes and expectations that adverse changes may occur (Zaleski et al., 2019). But it almost always involves an uncertainty about where and how one will be in the future, or how the future will match one's present preferences and values. People may also care about the future state of the world as such, or how their grandchildren will be doing. But even when they worry in altruistic rather than egoistic ways, they are likely to maintain an implicit reference to their own values and perspective. Experiences similar to future anxiety can be directed at the present, as when older people lament that it is "not their world anymore" and this makes them disapprove of it, even when it seems to exhibit plenty of progress and improvement (van Wijngaarden et al., 2015).

Before dismissing worries about preference and related personality changes as exaggerated or irrelevant, we should keep in mind that it is not just a question about what we have real reason to worry about. It is also a question of what might *feel* worrisome or bad, regardless of whether there is an objective basis for this or whether it can be said to be rational. People's tendency to cling on to their preferences and resisting seemingly beneficial personal changes is exemplified by a phenomenon famously highlighted by Proust in *Remembrance of Things Past*: we might experience emotional distress, for example, by suffering from jealousy, grief, or unrequited love, while at the same knowing very well that there will come a time when we no longer care about the person or thing that is presently causing us pain (De Sousa, 1987, 223). That is, we can predict our emotional trajectory and expect our suffering to end, and we might be able to think rationally about our situation, and adopt a more detached stance, which should enable us to employ the coping strategy of putting negative experiences in a larger perspective (see Section 3.7)—without this actually helping much. It is not unusual to actively keep unpleasant memories or painful experiences alive as a kind of reverence for the object of one's emotional investment, or this investment itself (and so, ultimately, out of respect for *oneself*, as one has been and presently is). Yes, one might experience some relief by adopting a more objective, detached attitude. But one's emotionally laden, temporarily and personally indexed perception of the situation presents it as being much more important than it appears from such a stance.

Transformative changes and the risk of self-alienation have been discussed in recent analytical existential philosophy under the heading *transformative experience*. Laurie Paul has highlighted the significance of important life choices that can effectively lead to becoming someone new, that is, to develop new values and preferences and gain access to new dimensions of life or experience. Her favorite example is that of becoming a parent. According to Paul, such a choice will be partly blind. It cannot be a completely rational choice, because one cannot know in advance which preferences and affective dispositions one will develop by becoming a parent, nor what it will be like in experiential terms (Paul, 2014).

Paul's analysis has been criticized and modified. It seems to exaggerate both our lack of epistemic access to future life conditions, like being a parent, and the self-alienation which it may engender. Many have objected that even transformative life choices need not be completely blind. We can to some extent rely on the testimony of others, and this need not alienate us from our choices (Carel & Kidd, 2020; Harman, 2015). Our own imaginative projection may also give us some knowledge about how it will be like to live a future life with different preferences and experiences (Kind, 2020). These corrections are justified. As I have emphasized myself, we are not completely in the dark neither as to how the future will be nor how we will ourselves experience it. Still, Paul aptly reminds us that we will ourselves change with new life circumstances, and that this change in values and preferences should be taken into account when accessing the new circumstances.

However, transformative experience as described by Paul does not quite fit the case of climate change-induced life changes. For such changes will not typically be brought about by conscious decisions. Some of them will simply befall the people involved. People might *expect* to change their preferences, without themselves having chosen to do so. This may arguably make the changes even more alienating than the transformations described by Paul, which may not be completely autonomous because of lack of epistemic access to the outcome, but which are still caused by a personal, conscious decision. Other changes may occur in ways that partly or indirectly influenced by choices and efforts of the individual, without being completely or directly caused by them. I may voluntarily, but only gradually, develop an interest in vegetarian cooking, which then leads me to seek information about vegetarian diet options; through inspiration and guidance found in the vegetarian literature, I may finally come to change my eating habit, without ever having consciously decided to do so. Being subject to nudging or "choice architecture" (Thaler & Sunstein, 2008; see also Section 6.1) is a further example of how people may come to change their behavior in ways that can be said to voluntary but are nevertheless shaped by the circumstances and even by the explicit intentions of others. Behavior change due to nudging will rarely be so significant as to count as a case of personal transformation. But it illustrates that there is a large variety of life change processes, and that they exhibit different degrees, forms, and mixtures of autonomy, agency, passivity, and coercion. Besides, even though nudging affects only one's particular choices and actions and not—as such—one's more fundamental values and preferences, it may contribute to changing the latter in the longer term. Being nudged to eat salad rather than meat can make eating salad a habit and so develop an active preference for it and a corresponding weaker preference for, if not aversion to, eating meat. Extensive changes in the choice architecture of one's environment may lead to quite extensive, if not profound, changes in one's habits and values, and so eventually also in one's personality.

Agnes Callard has proposed a modification of Paul's analysis of transformative experience which makes it less dependent on conscious choice and so more pertinent to the case of climate change. Apart from the transformative experiences or "revelations" highlighted by Paul, Callard notes, there are also transformative *activities*. These do not involve any radical break in the agent's biography, nor do they depend on any particular decision. They may therefore seem less autonomous. But arguably they are not. It is characteristic of the decision-based processes described by Paul that the subsequent changes in one's personality simply happen to one because of the decision. Once the dice is cast, the process is beyond one's control. By contrast, transformative activities as understood by Callard are *continuous processes sustained by the person's aspirations and efforts* (Callard, 2020). Examples of this are learning a language or learning to play a musical instrument. Little is achieved simply by forming the decision to do so, just as the decision to live more sustainably is not a good predictor of how one will actually live in the future. It is central to learning a language, learning to play a musical instrument, or living more sustainably that one keeps up the work it requires. The process is *transformative*, however, in that one cannot know in advance what it will be like to understand and speak Spanish or to play the piano; the latter will moreover likely change one's experience and appreciation of music. Transformative activities are expressions of *agency* not because they depend on a prior choice, but because of the agent's ongoing involvement and "investment" in the transformation. Calhoun speaks in a similar sense of *endeavoring*: to commit oneself to continuing a pursuit in spite of setbacks, obstacles, and feelings of disenchantment (Calhoun, 2018).

Transformative activities can be seen as a way of *becoming oneself*: changing, in a personally transformative way, while maintaining significant personal continuity, and without experiencing any harmful loss of self of or alienation; when embarking on a transformative activity, one is rather inclined to proleptically identify and sympathize with one's future self and life, despite having only an incomplete grasp of how it will be. This makes transformative activities attractive in the context of climate change: if we can get the personal and societal changes required by climate change to take the shape of, or align with, transformative activities, then this may make them less troubling and more acceptable (see Ch. 6). As we shall see, the notion of transformative activities also has close affinities with the classical notion of self-cultivation, especially when this is understood realistically (see Section 5.4).

Overall, we have seen that the concern for authenticity and conservation should not be taken too strictly, but that it is so strongly psychologically and culturally ingrained that it must not be ignored, either. We have also seen how it might be met in different ways and be compatible with quite significant changes in personality and life circumstances.

4.2 The Shape, Meaning, and Management of Life

Concerns other than that for authenticity might put constraints on what can count as good or satisfactory life changes. There may be independent value to continuity, which can, but need not always, be considered a form of conservation. Further global (or otherwise nonlocal, that is, temporally extended) features of a persons' life, like its meaning, might also be held to contribute to its prudential value. And although different ways of shaping one's life, of making plans and commitments, or weighing different aspects of it differently may have only instrumental value, this value is arguably so great and fundamental that they also deserve attention.

Let us first briefly consider the value of continuity and other possible "temporal shapes" of a life. What could be the value of continuity apart from the value we attach to preserving (and so remaining) ourselves? Again, while it is difficult to maintain that there can be any intrinsic value to continuity—after all, it is just one possible shape among others which a series of life events might take—it cannot be denied that we seem to value it intuitively and quite strongly. I suppose, however, that the value we ascribe to continuity is only relatively fundamental, as it derives from the more fundamental value of emotional states like *belonging*, *attunement*, or *unperturbedness*. We naturally value being secure, free of worries, not distracted, interrupted, or vulnerable. But this means that discontinuities that are not associated with discomforts of the latter sort should not be valued negatively, which they also are not, in practice. Leaps are welcomed if they are to the better.

Some might claim that there is something distinctively negative about living a life that "goes down," that is, living on a downward trajectory. If the number of good things one is able to enjoy is reduced, the negative value of this may be thought to exceed what can be accounted for simply by comparing the situation at different points in time; the very fact that one's life goes less well, its downward slope, might add to its badness (Velleman, 1991).

I do not think this claim can be sustained in its purest form. It fetishizes the shape of life, which is really nothing over and above the life whose shape it is. However, it should be acknowledged that the *perception* of one's life taking a downward trajectory—experiencing phases of decline or worsening conditions, or a simple reduction in the goods and options available—may indeed be an additional negative factor. This is an example of how the reflective and metacognitive powers of humans may exacerbate suffering. When we are not doing well, our awareness of this tends to add to our misery. However, this effect depends on the decrease or reduction being negatively valenced in the first place. Overweight people seldom experience weight loss as something negative, nor is a reduction of one's debts usually perceived as problematic. A reduction in wellbeing is per definition negative, of course. A loss of goods or options that are typically associated with, but not necessary for, wellbeing is also likely to perceived as negative, but need not be so. People react differently to having to curb their material consumption or do with fewer travel options and react differently under different circumstances. (We shall see, for example, how the social context matters (in Sections 5.3 and 5.5).) What is perceived as a harmful loss if it befalls only a single individual or certain group of people may not be experienced as such if the population in general is subjected to the same reduction. In this case, it may be assimilated to the "natural" changes that form a kind of inevitable background condition (see Section 2.4).

So much for the possible shapes and trajectories of life. Another potentially global feature of a life is the *meaning* that can be ascribed to it. Some might consider it a serious omission that I did not include meaning in my discussion of wellbeing in Chapter 3. For meaning is widely regarded either as an important factor in wellbeing or as a prudential value in its own right, which may compensate for a lack, or lesser degree, of wellbeing (Wolf, 2012).[2] Since we are, in any case, concerned only with prudential value, the relevant kind of meaning must be such that it can be said to benefit the person herself. A life can also have meaning without this making the person whose life is better off, for example, if the person does something that has tragic consequences for herself but turns out to be historically important. Yet this kind of meaning—meaning *of* a person's life, that is not manifested *in* her life and does not in some sense benefit *her*—is not in itself prudentially valuable.[3] Hence we should focus on meaning *in* life—a quality of the life in question *as it is lived*, a quality internal to it. But what is it? It is widely, and rightly, held to mean having *something to live for.* Meaning is purposiveness. Not any kind of purpose will do, however. It must be more than a mere reason for acting. Being on fire gives me a strong reason to extinguish the fire, and so a strong purpose, but it does not give me something to *live* for. The purpose in question must be something that is sufficiently great or good to make one's life *worth* living. Meaning in life is roughly the same as *worthwhileness.*[4]

A life with sufficient subjective wellbeing will, on this view, also be meaningful. But the reverse is not the case. Meaningfulness is no guarantee of subjective wellbeing. A person may lead a meaningful life, and experience it as such, although it cannot be said to go well. She may sacrifice herself for a noble cause, thereby suffering harm. There are many examples of artist, politicians, and ordinary altruistic persons of whom this seems to have been true. (Beethoven and Kierkegaard, for example, both described themselves as having chosen a life of unhappiness to be able to realize their artistic and intellectual potential (Beethoven, 1997; Kierkegaard, 1992, 512).) But while meaning in life does not necessarily make a person's life subjectively satisfying, it often does contribute to it. There are close causal connections between meaning and wellbeing. The experience of meaning in one's life may not suffice to compensate for the immense suffering and hardship one endures for the sake of a noble goal. But it will certainly mitigate the negative effects. Pains endured for a noble reason can be less bad than pains endured for a bad or for no reason. A strong sense of purpose can sometimes be sufficient to make one's life as a whole feel satisfactory, even though it contains many unpleasant experiences.

This is relevant to pro-environmental change. For although we should not follow proponents of green virtues or objective wellbeing in seeing pro-environmental attitudes and actions as directly conducive to wellbeing, they are likely to be so indirectly, due to their ability to make one's life and actions feel more meaningful. If one can come to experience one's efforts at behavioral change, and apparent sacrifices for the sake of mitigating climate change, as meaningful, this can be an effective way to maintain or even increase wellbeing during green transition.

The possibility of meaning without wellbeing remains relevant, however, and talk about sacrifice may be a reason for concern. The experience of meaning can be a source of genuine wellbeing and effectively mitigate displeasure. But it can also motivate action independently of this effect. People may feel motivated to do things they experience as meaningful even though this does not give them pleasure or make their life seem good to them. Although perceived meaning is still, in such cases, a piece of effective moral psychology and a potential factor in bringing about pro-environmental behavior change, it is hardly a prudential value. It could be relevant to proponents of the view that responding to climate change requires genuine sacrifices and must be driven by moral concerns (that is, of view 2) in the taxonomy in Section 1.4, as it might make it seem more realistic.

But it is less attractive to those who, like me, would like to think that the necessary behavior changes can feel satisfactory or even gratifying.

The very focus on global features of a human life may carry a risk of sacrificing wellbeing for the sake of something else, just like the concern for authenticity can urge people to stick to their identity when it would be better for them to change it, holding them hostage to a life that is not really good for them. It is common to distinguish *lifetime* wellbeing from *momentary* and *periodic* wellbeing (Bramble, 2018). Some have argued that momentary and lifetime wellbeing are two irreducible kinds of prudential value and so can be the object of competing interests. Perhaps most famously, Velleman has claimed that a person has both a synchronic and a diachronic identity, each with its distinctive interests; we live our lives both from the successive viewpoints of individual moments and from a comprehensive, diachronic point of view (Velleman, 1991, 67).

I think this is another example of view that cannot be maintained in its purest philosophical form but may nevertheless highlight psychological phenomena with importance for wellbeing. Velleman's bifurcationism implies that we have radically split selves or dual personalities. This seems implausible and unnecessarily radical. It is more plausible to maintain to that we—as more or less unified, numerically identical selves—are able to shift our attention between the present moment and more global features of our life, and that we can give different priority to the two perspectives. Besides, the idea that momentary and lifetime wellbeing are two different kinds of intrinsic value, with no obvious way to resolve the resulting conflicts of interest, puts the very notion of prudential rationality in jeopardy, since it makes it difficult, if not impossible, to determine what is in our best interest (Bramble, 2018).

It is a psychological fact, however, that the two perspectives can, and often do, come into conflict, and that this can seem like a tension within oneself. People might feel themselves to be victims of their own larger-scale projects or, conversely, despise of their momentary or periodic pleasures. I remember myself damning my athletically ambitious self during track races for having gotten me into this pitiful state, underestimating the pain I would have to endure to achieve what now seemed to me a not so important goal. I also remember myself coming to view some of these races as significant accomplishments and as tough, but good experiences overall, once they were completed—usually only seconds after having crossed the finishing line. Was my running self a victim of my ambitious and pain-neglective planning self, or was it itself neglective of the greater good of athletic achievement? Being committed to a holistic view of wellbeing (see Section 3.10), I tend to side with my planning and achievement-enjoying self, which also seems to me to put negative experiences in a larger, but appropriate perspective, thus using an efficient and legitimate form of hedonic adaptation (see Section 3.7).

But although a sensible holism about wellbeing does oblige us to look to the temporally extended features of people's life, this point should not be overstated. Again, it is about achieving an appropriate balance. As pointed out above, there is also a risk of fetishizing global features of one's life, and one can indeed become a victim of one's own projects and plans if their realization causes to much suffering and does not sufficiently mitigate or compensate for it. Brainwashing or religious or political fanatism seldom makes for really satisfactory lives. Lifetime wellbeing should still be seen as depending, at least for the most part, on the wellbeing one enjoys during shorter periods of time.[5] Meaning, self-understanding, projects, plans, and commitments can be conducive to a person's wellbeing, but only insofar as they contribute to make her life feel more satisfactory. We might indeed reduce the harms brought about by lifestyle changes needed for green transition by experiencing them as meaningful and enjoying our success at becoming better persons and doing good. But this requires us to psychologically internalize the values and projects to such a degree that they do not merely complement but genuinely mitigate the harms in questions.

4.3 Temporal Styles and Agents of Change

People are different. My process-oriented approach to wellbeing is motivated by a recognition that what is good—or best—for a person in a specific situation may depend strongly on the kind of person she is and where she is in life. There is much truth in the classic Socratic maxim "know thyself," though it should be complemented with a more general prescription to be aware of people's particularities when caring for them. People have different emotional selves—different dispositions to react emotionally. They also have different ways of imagining and relating to the future (and different ways of relating to the past and the present, as well). Calhoun has pointed out that people have different *temporal styles*—different ways of managing their diachronic existence. Some care strongly about long-term goals and the overall character of their life, taking on whatever task seems to be necessary for achieving these overarching aims. Others are more prone to change their plans if they encounter obstacles or the actions needed to achieve their long-term goals turn out to be less rewarding (Calhoun, 2018).

This distinction matches the observation above that one can care more or less for present or overall wellbeing. It is relevant for our purpose in two central respects: First, it indicates that life changes may affect people's wellbeing differently, depending on their temporal style (which thus turns out to be a further mediating factor in wellbeing). Long-term committers may suffer more from having to adjust their ways and manners; people living in the now may have an easier time doing this, but also feel more negatively affected by immediate tasks and problems. In short, the long-termers are sensitive to disruptions of their life-course and identity; the short-termers are more sensitive to a change in the content of their life.

Second, it may be easier or more difficult for people to commit themselves to long-term projects and carry on with them in the face of resistance and unanticipated problems, depending on their personal, temporal style. Long-term committers may not be prudentially superior; they may not actually get the best result for themselves out of sticking to their plans and overall goals. Calhoun argues persuasively that committing oneself to projects is not necessarily better than merely intending to carry them out, and so be more open to reconsideration. She also points out that even if one gives special value to coherence or narrative unity of life, long-term commitments are not the only vehicle for securing this. As she remarks, "lives marked by many changes in careers, relationships, ways of life, and social identities can have a high degree of narrative unity and integrity" (Calhoun, 2018, 103).

However, long-term committers are likely to be the most effective *agents of change*. Assuming that it is psychologically possible (see Chapter 5) and morally acceptable (see Chapter 6), we may thus have reason to cultivate such a temporal style. (Calhoun does not herself say anything about whether temporal styles are fixed personal traits or can be modified. But it is reasonable to assume that like most other human behavioral patterns, they are biologically and socially conditioned, but still to some extent malleable.) We may even have some prudential justification for doing so. For it can be argued that even though long-term committing may not generally make one better off, it may be the best way of adapting to the current conditions. If we are now required to care more strongly about long-term goals—greenhouse gas reduction, for example—for moral, political, or pragmatic reasons, then the long-term temporal style may make it easier (more pleasant, less frustrating, etc.) to comply with these requirements and societal trends. This points to the importance of personality formation or *Bildung*, which shall occupy us in Chapter 5.

4.4 Keeping the Good Things Alive—Variation as the Key to Wellbeing?

People may seem to prefer continuity, but they also like variation. A good life is not supposed to be monotonous. It is not unusual to ascribe intrinsic value to variety, diversity, and living dynamically. Liberalism, with its emphasis on a large option space, seems committed to this idea, as do theories of human flourishing. Wilhelm von Humboldt, who not only contributed to the idea of self-formation or *Bildung*, but also formulated the anthropology and value theory implicit in modern liberalism, viewed diversity and variation as the essential conditions for a good human life. In an early diary note, he stated that "[t]he principle that one must have been in many situations of every kind is so well established with me that any new situation is agreeable just because of its newness" (Humboldt, 2015, 69).

This view may, again, not be ultimately justifiable. Just as with continuity, the formal properties of variation and change may not have any intrinsic value, independently of their specific content or manifestations. There may also be psychological limits to how much change and diversity people can cope with. But monotony causes boredom and dissatisfaction, and various kinds of movement, variation, and diversion are known to be important sources of joy and gratification. We have already encountered an argument for the pro tanto value of diversity of experiences and activities: the phenomenon of diminishing marginal utility implies that one will get more out of pursuing new and different interests or experiences. Even when one might get more out of deepening an existing interest or familiar experience, this also involves seeking variation. It requires one to register further nuances within an already familiar field and so to appreciate its internal variety.

Climate change and our responses to it threaten both to make life too simple and dull and to cause stressful changes. The human predilection for diversity requires us to ensure that vegan, vegetarian, and low-meet food options are also sufficiently varied. We already know this is possible. But we also know that possibilities are often far from realized and widely acknowledged. People are forced to eat less sustainably because of a lack of easily available or visibly attractive options. Similarly with clothing and other consumer goods. It is possible to produce and distribute fashion sustainably and still satisfy the human desire for variation, decoration, expressivity, and exploration. But as things presently are, this requires costumers to put considerable efforts into finding and becoming accustomed to the relevant options. As such, we seem to be threatened with both dullness and stressful change at the same time: we might fear having to do with less fancy, less varied, less fun and convenient products and leisure activities, but also having to alter our consumer and other everyday habits considerably.

A related worry is that sustainable living might indeed be sufficiently challenging and diverse, but that we will experience the wrong kinds of challenges and diversity. Vegan food is already quite diverse, and putting together a satisfactory diet requires quite some browsing and experimentation—and, some might say, this is rather a source of irritation. Better just go for a familiar non-vegan dish, which gives you a rich culinary experience without having to fight your way through a jungle of different, only half-satisfactory alternatives. We wish to explore, discover, and make surprising findings and bargains, not to search for a needle in a haystack.

Whether or not such worries are justified, it seems that we need to strike a balance between convenience and familiarity on the one hand and challenges and variation on the other. Changes should be large and significant enough to stimulate one's curiosity, but no so large or radical that they cause fear, frustration, inconvenience, or loss of control. Here we should remind ourselves that competence, a feeling of mastery and ability to operate effectively within one's life context, is presumably a fundamental human need (see Section 3.6). This indicates that changes and challenges are perceived most positively when they are embedded in a larger life context which one does feel able to master and control. This context need not itself be static or simple, but it should remain manageable and afford sufficient safety and attunement. The saying that variety is the spice of life seems apt, since the variety must, like spices, be properly dosed.

The importance of variety for wellbeing has been demonstrated empirically. It has been shown to be an efficient means to prevent negative hedonic adaptation. In Section 3.7 we were concerned with positive hedonic adaptation, that is, the tendency to come to terms with worsening circumstance and so maintain wellbeing in the face of adversity. But people display a similar tendency to adapt to positive life events, so that their positive impact wears off. This is presumably a driving factor behind the tendency to want still more and seldom remain satisfied with what one already has. Hence preventing this form of adaptation from occurring, or mitigating it, is also an important means to achieving wellbeing in a sustainable manner.

Positive events fall out of a person's awareness if they are not manifested in varied ways (Lyubomirsky, 2011). Greater variety in people's positive events has been shown to temper aspiration and so allow them to remain satisfied with their life situation (Sheldon & Lyubomirsky, 2012). *Surprising* positive events have been shown to be particularly resistant to adaptation (Wilson & Gilbert, 2008). They are difficult to explain away as just being part of the normal, everyday conditions; they keep things "exciting and fresh" and so make it easier for a person to remain content with what she has (Armenta et al., 2014). The importance of keeping things fresh and preventing them from becoming humdrum can also be seen from the way people reacted to the recent COVID-19 pandemic lockdown. During the first wave, especially in the initial phases, many seemed to perceive the new situation as an interesting change with opportunities for new positive experiences; during the second wave, the novelty wore off and the negative and monotonous aspects of life during lockdown became more salient (Łaszewska, 2021; Mau et al., 2022).

In a recent book that expounds and celebrates the idea of regaining sensitivity to the good things in life by seeking variation, Sharot and Sunstein claims that "one cannot habituate to change" (2024, 23). This is vivid and roughly fitting way of emphasizing the importance of variation for wellbeing and its power to avoid the negative forms of hedonic adaption. It is also slightly exaggerated, however. We have seen in Chapter 2 that certain changes come to be perceived as background conditions, that is, as normal and expectable. And one can—and must—learn to cope with change, and often do so by getting more used to them.

The empirical research on the impact of variation on wellbeing focuses primarily on the quality of external events and stimuli, like their objective variety or random character. The impact of these qualities is an instructive example of how external conditions matter for wellbeing in ways that are relatively independent of differences in personal dispositions or subjective uptake. Surprising events can pierce through the shield of habits and expectations and make an impression on persons that are otherwise not prone to viewing things particularly positively. (It should be kept in mind, though, that the same holds for surprising negative events, that likewise retain their negative impact for a longer period; this is of course also relevant for the prospect of climate change.) The empirical research thus seems to confirm that the presence of certain goods or objective factors can be a reliable indicator of wellbeing (as I conceded in Section 3.8). We should not, however, forget that variation and surprise are highly abstract properties, and that the empirical research does not indicate that variation of any specific thing is required for it to make a positive difference.

More importantly, we should acknowledge that variety and even surprise can also be achieved—or intensified—through subjective attitudes and strategies. The empirical research concentrates on vivid manifestations of novelty and change. But it is also possible to be particularly attentive to diversity and variation and so to achieve a similar effect even though little new happens to the external stimuli. This part of the strategy of cultivation we considered in Section 3.9, and which was recommended by Kierkegaard. As for surprise, it seems natural to think of this as a matter of chance and thus something that can only come from without. But although you cannot, of course, become surprised by something you already knew or planned, it is possible to pursue interests and courses of action that make it more likely that one will experience surprise. Maintaining open-mindedness,[6] curiosity, and an experimenting and not too risk-averse attitude to life contribute to this. Again, there is a trade-off, because such an attitude also increases the risk of experiencing negative surprises. Considering the importance of some kind of surprise for wellbeing, however, and how difficult it is for most persons to do well without some element of noticeable change and variation in their life, it may well be a risk worth taking.

Finally, although it may seem that climate change and our response to it promise only boring monotony and negative surprises, it could well be the other way round. We should remember that the qualities that support wellbeing most widely and reliably, like change and variation, are highly generic. It is an open question by what or how they are instantiated most efficiently in terms of their impact on wellbeing. Just as one should not be too quick to judge one cultural or societal change as being larger than another, because significant changes can be subtle and materially extensive changes societally insignificant (Section 2.4), so one should not make judgments about the monotony or diversity of situations simply by counting items and options. Less can turn out to be more, and more can turn out to be less. Contemporary consumer culture is arguably rather bland and uniform, if not "sensually impoverishing" (Soper, 2020, 66), at least in its most mainstream varieties. Extensive option freedom does not have the gratifying and vitalizing effect Humboldt assumed it would have. Options may be irrelevant or practically unavailable, yet people's awareness of them may provoke a sense of missed opportunities or societal pressure to conform to unrealistic expectations. This may in turn cause stress and anxiety (Schwarz, 2004) and prevent people from more intensively cultivating and appreciating the options at hand.

A good example of how a seeming change to a less dynamic and thus duller way of life may in fact turn out to be quite the opposite is to start cycling instead of going by car. It is objectively slower but may nevertheless better satisfy one's need for speed. Cycling makes "riders subject to rhythms that they themselves create and sustain" (Ryle in Ryle & Soper, 2002, 102), thus also better meeting the need for felt competence and autonomy; it "gratifies the interest in going faster while remaining ecologically benign" (Soper, 2020, 112). Another example is reusing products, which will likely force us to become more engaged in searching and selecting, and which may hold more surprises than going for the brands you already know or merely browsing through company catalogues. As one very insightful lifestyle blogger remarks:

> A second-hand store is like a box of chocolates. You never know what you're gonna get. There's so much variety in a single store that it's an experience in itself. You get to compare quality, search for lost treasures, learn what different styles have to offer, and light up your creativity. All in one place. Still think second-hand is not for fashionistas? [...] The fast industry designs your consumer habits, having you always feel out of trend and making you buy more and more clothes. Although this gratifies your desire for change and excitement, you gain no long-term satisfaction. Second-hand stock changes every month or so, and allows you to browse endlessly. And no store is the same.
>
> (LisaLisa, 2019)

There is even a possibility that the larger societal transformations caused by climate change will be experienced not as stressful ruptures, but also as enlivening variations in life conditions and circumstances. This may sound like I am romanticizing climate change. But that is the last thing I want to do. The realistic stance I am advocating requires us to not downplay potential harms and even recognize that seemingly small changes can be perceived as quite uncomfortable. However, it should be remembered that the impact of societal transformations is also conditioned by attitudes that may to some extent be cultivated, and that societal changes can be managed, orchestrated, and communicated in ways that make them easier to come to terms with, if not genuinely exciting. They may also stimulate collective actions in ways that boost community cohesion (Adger et al., 2022), as well as personal engagement.

4.5 Values and Impacts of Personal Change

Personal development is widely regarded as highly valuable, sometimes as simply essential to a good human life. We tend to dislike not merely lives in which new happens, but also lives that do not themselves exhibit some kind of significant change. People may be averse to downward trajectories, but a flat life trajectory may seem just as bad, if not worse. Let us look more closely at the putative value of personal change.

Most accounts of the value of personal development focus on its consequences—the personality and life that can be expected to result from undergoing the changes in question. Personal development is considered good for people because it makes them more capable of coping with difficulties, getting what they want and appreciating the good things in life. The changes are thus seen as a means to an end, viz. to become a person who will probably live a more satisfactory life.

Yet personal development can have value other than by fostering habits and abilities that may prove useful later in life. People get satisfaction out of the attainment personal goals. Having improved oneself and realizing that this is so can be a source of more immediate gratification. Attainment of personal goals has been shown empirically to support positive changes in meaning in life and wellbeing (Messersmith and Schulenberg, 2010; Sheldon, 2008; Stauner, 2013). This is particularly so when the goals are intrinsically motivated and there is a lack of environmental support (which is itself conducive to wellbeing but makes actual goal attainment less important). The positive effect on wellbeing of goal attainment is also strengthened by a person's willingness to invest and her overall self-determination (Stauner, 2013). This further indicates that the experience of striving and succeeding is important to wellbeing. Sharot and Sunstein summarize recent empirical findings by noting that "[j]oy often comes from perceiving yourself as moving forward, changing, learning and evolving" (2024, 23).

That there may be a positive quality to undergoing personal changes, apart from that of the goals or other consequences, has been emphasized by both the tradition of self-formation or *Bildung* and existential philosophy (though the former also acknowledges the instrumental value of personal development). The two traditions intersect most strongly in Kierkegaard, who describes quite elaborately how facing and meeting the task of self-improvement can itself be a source of considerable gratification. In the second volume of *Either/Or*, Kierkegaard lets his protagonist, the self-declared "ethicist" Judge Wilhelm, explain this both abstractly and with examples from his own life. Judge Wilhelm notes, first, that the objective accomplishments of his work are not essential; it is the very willing and striving for improvement that matters:

> I therefore disclaim all that self-importance that calls so much attention to itself in life; I do my job and do not waste my time calculating whether I accomplish anything. So what I accomplish as a result of my work is my good fortune, in which I might well dare to rejoice but which I do not impute altogether to myself.
>
> (1992, 568f.)

Wilhelm goes on to describe how enacting his role as a judge, husband, and father is a primary source of joy and gratification:

> [...] I perform my services as judge, I am glad to have such a vocation, I believe it is in keeping with my abilities and my whole personal being, I know it makes demands on my powers. I try to mould myself more and more to it, and in doing so I feel that I am developing myself more and more. I love my wife, am happy in my home. [...] My work has importance for me, and I think it also has some for others, even though that is something I cannot determine and exactly measure. I feel joy in the fact that the personal life of others has importance for me, and desire and hope that mine also has some for those with whom I am in sympathy in my whole view of life ... Thus I love existence because it is beautiful, and hope for one that is more beautiful still.
>
> (1992, 582)

Like Kierkegaard's treatment of cultivation in *Crop Rotation* (see Section 3.9), this apparent apology for self-development should not be taken at face value. It is also somewhat ironical. Like Humboldt, who warned that even the attempt to form humanity and foster personal growth could itself degrade into monotony (Humboldt, 2000, 61), Kierkegaard is aware that the pleasure of self-development can turn into smug self-satisfaction and the development itself becomes an act of conformity and routine. Yet he is still describing a genuine and important kind of pleasure. Central to Kierkegaard's view is the notion that the pleasure of trying to improve oneself is in large part aesthetic. When one lives in an active and committed manner and is aware of doing so, one experiences one's own life as beautiful. This may sound strange and exaggerated. But the idea is not to subject one's life to aesthetical evaluation. It is that one can experience it as a pleasant, harmonious combination of freedom and necessity, general and particular, and tension and resolution—much like we would experience a work of art or beautiful landscape.

More generally, Kierkegaard lets his protagonist describe how taking one's duties and task, and doing so wholeheartedly, gives rise to a feeling of vitality or vigorousness, in an existential, rather than merely bodily sense. This is similar to what Haybron (2008) describes as the fundamental emotional state of *engagement*. The latter may also cover more world-directed activities not aimed at self-improvement or self-realization. However, when such activities are pursued in a way that is sufficiently engaged, that is, with sufficient identification and investment, they are often, if not always, accompanied by at least an implicit sense of realizing oneself. Vendrell-Ferran (2021) has argued, similarly, that feelings of vitality are forms of self-experiencing. It should also be noted that liveliness or vivacity is generally considered an attractive quality in others, something that makes them more agreeable and more pleasant companionship. While Kierkegaard is focusing exclusively on the subjective or "inner" qualities of action and self-improvement, as is typical of existential philosophy, its outward manifestations may also be prudentially valuable (if not intrinsically so). Being with engaged and vivacious people feels good and is likely to be socially and emotionally contagious (Decuypere & Schaufeli, 2020).

Existential philosophy has also emphasized the value of profound personal change or existential conversion, notably the act of "choosing oneself" in the sense of unreservedly accepting oneself and wholeheartedly taking on the task of living one's life. Although this is intended as a description of something more fundamental, it seems similar, at least in certain respects, to some of the adaptation and coping strategies highlighted in Section 3.7. For example, identification with one's condition is known to be an efficient means for coming to terms with otherwise adverse circumstances; the same holds for adopting an appropriate perspective.

One important similarity is that these are all changes in personal attitude and outlook that leave the external circumstances unaltered. Wittgenstein underscores this point in his existential-philosophical meditations toward the end of his *Tractatus*, where he claims that "if good or bad willing changes the world, it can only change the limits of the world, not the facts [. . .]. In brief, the world must thereby become quite another. It must so to speak wax or wane as a whole. The world of the happy is quite another than that of the unhappy" (Wittgenstein, 1922, §6.43). Here, the adoption of a new and fundamentally different view of the world is described as a *gestalt shift*: an abrupt change that pertains to everything one is confronted with, but without any change in this itself. The stimulus remains constant, but its meaning becomes completely new.

A vivid example of an existential conversion, and the way it is assumed to be both life-changing and—mostly—practically inconsequential, is provided by Tolstoy in *War and Peace.* His protagonist, Pierre, has long been depressive and disoriented, lacking and vainly seeking purpose and meaning in life. Having faced execution and experienced both atrocities and extraordinary human kindness during the sacking of Moscow by the French in 1812, Pierre adopts a completely new outlook. Tolstoy describes him as having changed fundamentally, although he remains very much himself:

> Outwardly Pierre had hardly changed at all. To look at he was the same as before. He was just as absent-minded as he had always been, and he seemed to be permanently preoccupied with something that wasn't there, something that was all his own. The difference between his former state and the one he was now in was that [...] [n]owadays he could still be oblivious to everything that was going on around him and what was being said, but at least he looked at what was going on around him with the ghost of a smile, however ironical, and he also listened to what was being said, though he was obviously seeing and hearing something very different. In the old days he had seemed like a nice man who was unhappy, which inevitably kept people at arm's length. Nowadays a smile of joie de vivre played constantly about his mouth, his eyes shone with sympathy for others, wondering whether they were as happy as he was, and people enjoyed his company (Tolstoy, 2016, 1230). [...] In practical affairs Pierre suddenly sensed that he now had a centre of gravity that had been missing before. Before, every monetary question, especially requests for money, which came his way all the time because he was a wealthy man, had always left him worried and perplexed [...] Nowadays, much to his own surprise, he found he had no more doubts or misgivings over any such questions. Now there was a judge within him determining what he must do and not do according to a set of laws of which he had no understanding.
>
> (Tolstoy, 2016, 1233f.)

Tolstoy is keen to emphasize that having found a "centre of gravity" does not imbue Pierre with special powers. It does not even cure him of his absentmindedness, and he remains socially awkward. On the other hand, Tolstoy does not describe his existential conversion as completely inconsequential or without practical benefits. For one thing, Pierre is now happy. He is also more tolerant of others, less worried, and in some respects more resilient (though the subsequent narrative reveals that there are still limits to what he can put up with; he has not become completely imperturbable). Besides, people now enjoy his company. So it seems, if we are to believe Tolstoy's description, that not only vivacity and engagement but also existential integrity can be perceived as pleasant and likewise be a social asset.

I have expressed skepticism about the extent to which people are able to acquire a new mindset, still more about the possible impact on actual behavior on such a shift. Existential philosophers and likeminded authors like Tolstoy do seem to contradict this, inasmuch as also they emphasize that people are likely to keep on doing things they used to do even after acquiring a new integrity or seeing the world through a different lens. Yet they also indicate, albeit to different degrees, that such changes can make *some* real difference. If we allow ourselves to interpret their view more concretely psychological, we might say that it ascribes various kinds of potential indirect impact to existential conversions, for example, due to shifting awareness and changing orientations and values. Such indirect effects are not implausible, but also presuppose that the person in question has really changed her fundamental outlook and fully internalized it. There is an important difference between the kind of change in mindset we are concerned with here and those usually considered in the environmental humanities. For in the existential case, the change does not consist in the acquisition of a new set of beliefs. It consists more literally in the adoption of a new mindset, a new way of seeing and approaching the world as whole; and it requires full identification with this mindset, making it into a way of being rather than merely a way of thinking. Tolstoy describes Pierre as having formerly held many noble and assumedly useful beliefs, but without having ever really appropriated them or taken them to heart. By contrast, when Pierre finally manages to unreservedly identify with a certain outlook, the precise content of this seems to matter less. Pierre himself is not able to articulate it, and Tolstoy also does not say exactly what it amounts to, though it seems to involve commitment to a form of humanism, a religiously tinged humility, and endorsement of the world as it is, as well as fundamental self-acceptance.

Precisely what the kind of strong identification emphasized by existential philosophy amounts to, psychologically speaking, calls for further clarification. Contemporary moral psychology distinguishes between merely *avowed* and properly *internalized* norms. But internalization is typically understood differently, as being rather a matter of social conditioning. Kelly defines internalized norms as "socially acquired behavioural rules stabilized by communal practices of intrinsically motivated compliance and enforcement" (Kelly, 2022, 6). This notion is also relevant for our purpose, as it describes an important phenomenon that has otherwise not been sufficiently acknowledged. We shall revisit it when we come to consider the role of social influence and peer pressure (Section 5.5). But the existential philosophers seem concerned with a kind of internalization that is neither brought about nor maintained by social factors.

As far as I know, the potential impact of fundamental self-acceptance (or other forms of deep personal change) on mastery of life has also not been studied empirically. It also appears difficult to do so, because the phenomenon we would like to study is notoriously elusive. However, studies of the role of self-determination and goal pursuit, to some extent also studies of the role of meaning in life, seem to have been targeting related phenomena. The constructs employed in these studies can perhaps be seen as ways of concretizing and operationalizing the lofty philosophical notions. These studies seem to confirm the assumption that deep attitudes and commitments do matter for wellbeing and mastery of life. This has potential relevance for green transition, since it indicates that lifestyle changes may be easier and feel less uncomfortable, and be pursued more persistently, if a person identifies strongly with her overall aims and projects.

However, although engagement can do good and be put to use in green transition, we should keep in mind the importance of a balanced life and the fact that wellbeing can be achieved in a variety of ways. It may now sound as if a life defined by stern commitment and incessant activity were the sole key to wellbeing. Yet of course it is not. Calhoun is right to emphasize that people are different, also with respect to which attitudes to life and time that are best for them. Some may have less need for self-determination and commitment, for example, if they are more naturally attuned to their life circumstances and able to just do whatever they need to do, without any need for dramatic conversions or confrontations with themselves. And almost everybody needs idle moments, time that is genuinely free, also of the obligation to improve oneself. Some have gone so far as to even postulate a need for *boredom*. Russell claimed it to be an acceptable side-effect of avoiding too much excitement, which "dulls the palate for every kind of pleasure," thus linking it to the need for cultivating experience and keeping it new and fresh (Russell, 1930, 62). In a similar vein, Elijah Millgram has suggested that boredom has the function of every now and then creatively destabilizing the self, pushing one past the structures of ends that one might otherwise have remained stuck to (Millgram, 2004).

Fortunately, as we have seen (Section 2.6), life under and after green transition may also hold opportunities for more genuine leisure time. We may have to engage in combating climate change, taking on mitigation and adaptation task, and parts of everyday life may require more efforts on our part. But we may also become freed of the necessity to produce and consume ever more commodities. As the anthropologist David Kaplan has noticed, being a consumer in contemporary society is *work* (quoted in MacKinnon, 2021, 49), and we might become relieved of some of this work if the pressure to consume and the options for consumption are reduced.

However, the very opposition between activism and idleness should be taken with caution. Activities are different, and so are times of apparent inactivity. There are ways of getting nowhere in life, or having nothing to do, that are just plain bad. When Kierkegaard describes boredom as a fundamental evil, he takes it to signify a lack of meaning in life, conceiving it as an attitude of indifference which makes one incapable of any real joy (Klausen, 2020a). Russell, who is aware that there are different types and senses of boredom, is concerned with its more superficial manifestations, like a temporary lack of things that calls for one's attention. Similarly, when the existential philosophers emphasize the value of engagement and activity, they are not thereby propagating an "active lifestyle," still less making an apology for taking on ever new tasks and projects. Kierkegaard was highly the "busyness," which he saw as characteristic of his time (1992, 46), and he and Heidegger (2010, §36) relate boredom to idle curiosity and diversion, a connection exemplified by Tolstoy's protagonist Pierre before his conversion. The existential philosophers present existential engagement as a contrast, and possible antidote, to being constantly diverted and tempted into pursuing further interests and projects that do not resonate with themselves.

More generally, we should probably distinguish *work* in the narrow sense, as the carrying out of an externally motivated, necessary task, from *activity*, understood as an internally motivated and autonomously executed process. Or we may avail ourselves of the still finer distinctions made by Arendt, who distinguishes *labor*, what we have to do simply in order to meet our material needs (that is, what is necessary for consumption), from work in the sense of producing something more enduring, and *action*, which expresses human freedom and discloses the identity of the agent (Arendt, 2018). These distinctions are relevant to green transition not only because they indicate how a societal transformation that leads to less labor and more work might increase people's wellbeing, but also because they entail that it will make life more sustainable, thus forging a link between the concern for wellbeing and environmental issues. The following passage from Arendt's *The Human Condition* now reads like a remark about sustainable clothing ("sustainable consumption" would be a contradiction in terms for Arendt), though it occurs in the context of a meditation on fundamental philosophical anthropology:

> If one construes, for instance the nature of use objects in terms of wearing apparel, he will be tempted to conclude that use is nothing but consumption at a slower pace. Against this stands [...] that destruction, though unavoidable, is incidental to use but inherent in consumption. What distinguishes the most flimsy pair of shoes from mere consumer goods is that they do not spoil if I do not wear them, that they have an independence of their own, however modest, which enables them to survive even for a considerable time the changing moods of their owner.
>
> (Arendt, 2018, 138)

Arendt reminds us that there is a difference between the fact that nothing human-made, even that which is most carefully produced and used, lasts forever, and the assumption implicit in throwaway cultures that things are made not to last but only to be consumed.

Arendt's distinctions are illuminating and useful, and we should generally distinguish different kinds of activity when trying to pass judgment on their value and disvalue. But we should also be open to the possibility that a certain kind of activity can become endowed with a new meaning and so take on a new value. This is entailed by the subjective wellbeing approach. It also seems to be supported by Kierkegaard's existential analyses. What Kierkegaard illustrates with the self-description of judge Wilhelm is that one *can* come to identify with one's work and duties to such an extent that they do not feel like labor at all, and that they bring forth something new and disclose the identity of the agent, just like action does according to Arendt. Thus action and work, maybe even action and labor, can come to merge. This is an important coping strategy and again points to the integrative process of self-formation or *Bildung*, to which we will turn in the next chapter. It does not rule out, however, that the task of identifying with one's work or labor can itself become a stressful duty, or an obligation imposed from without. This danger is also implicit in Kierkegaard's description of the judge, who at times seems overly eager to play his role as a civil servant and family man rather than really being himself.

The same activities can thus take on different values and roles. They can be leisurely or stressful, sometimes an intricate combination of both. They can resonate with, but also feel alien to the person who acts, even when they seem objectively meaningful.[7] What Arendt describes in high prose as action may not be experienced as an equally rewarding or self-fulfilling by all. A relevant example of the shifting and sometimes ambiguous values that an activity can acquire is that consumption may indeed have become work but is also pursued as a popular means for diversion and relaxation. MacKinnon remarks that "[a] day at the Mall [has become] as much a family outing as a day at church or the ballpark" (2021, 52). And it is pursued as a means for relaxation not least because it is far from static, but involves activities of seeking, discovering, and risk-taking ("should I by this? Does this really suit me?"), thus merging feelings of safety and freedom from external pressure (though these may be largely illusional, according to MacKinnon and other critics of consumer culture) with feelings of excitement, of being challenged, and of action and resoluteness.

That one can feel almost serene while being engaged in outwardly dynamic and challenging activities is further confirmed by the psychological theory of *flow*. This state, which is assumed to be among the most enjoyable of all, is usually described as one of being completely immersed in an activity or engrossed in one's work (Csikszentmihalyi, 1990). Yet characteristically, flow theory does not distinguish between physical or mental activities and does not rule out that one can experience flow while doing work or even something that might be described as labor. One might become absorbed in activities aimed only at producing goods for immediate consumption, like preparing a simple meal.

Hence we must once again conclude that wellbeing can be achieved in a plurality of ways. Activity and engagement, especially engaging in and experiencing one's own development and improvement, can be a distinctive source of wellbeing, and it is compatible with the need for temporary freedom from obligations and commitments. But there is also a risk that the task of self-development becomes an unpleasant burden if one does not properly identify with it.

This concludes my treatment of the various aspects of living with and through change. We have seen that the concern for authenticity runs deep, but that we ought not be particularly worried about losing ourselves in the green transition. We have seen that people can value and deal with time differently, as some is focused on the present and others more prone to make long-term commitments.

We have noticed the importance of different kinds of variation for wellbeing: lifestyle changes and climate change interventions have a better chance of succeeding if sufficient diversity can be maintained. We have moreover seen that living with engagement and commitment holds its own pleasure. Equipped with these insights, including an emerging understanding of the potential of the idea of personal formation or *Bildung*, we can now face the tricky issue of how to make lasting behavior changes.

Notes

1 A vivid example is the acclaimed novel *A Fortunate Man* ("Lykke-Per") by the Danish Nobel Prize winner Henrik Pontoppidan, whose protagonist eventually abandons his, seemingly successful, quest for happiness to live—and die—in apparent misery, but in concordance with his innermost nature.

2 When meaning is treated as a prudential value in its own right, wellbeing is usually understood hedonistically, as being roughly synonymous with happiness.

3 It is a matter of philosophical controversy how far objective features of a person's life that are not perceived by herself as meaningful (or otherwise appreciated or enjoyed) can nevertheless add to its meaning in a prudentially significant way and so be said to benefit her or make her better off. Objectivist views like that of Metz (2013) imply that they can. I will largely ignore this possibility, however, and assume that meaning in the relevant sense must make some subjectively manifest positive difference to count as a prudential value at all (but still briefly consider how the objectivist view could apply to the case of green transition).

4 While this is treated by many as almost a platitude (e.g., Solomon, 1993; Haack, 2002; Wolf, 2012, 2), the identity of meaning in life and worthwhileness has been questioned by Metz (2012). However, Metz's argument depends on the controversial assumption that happiness cannot be sufficient for meaning (as it is hard to deny that it could suffice to make a life worthwhile), and it has been convincingly criticized by Matheson (2020, 313–24). https://doi.org/10.1007/s11406-019-00082-8.

5 This view has recently been criticized by Bramble (2018), who argues that there is really no such thing as temporal, but *only* lifetime wellbeing. I agree with many of Bramble's specific points, for example, that experiences necessarily take time and that this calls into question the idea of literally momentaneous wellbeing. I am also sympathetic to his view that it is only lifetime wellbeing which can be said to have fundamental, intrinsic value and so determines what prudence requires

of us (though I take it that this is just a limited version of the still more general point, known from utilitarianism, that what is best overall must be determined by no less than total (and ultimately person-independent) wellbeing). Yet in contrast to Bramble, I do think that episodes in a person's life may also be considered intrinsically valuable and that lifetime wellbeing is, at least partly, constructed out of such temporal wellbeing.

6 In the philosophical literature, open-mindedness has been primarily conceived as an epistemic value, consisting in a disposition to reconsider beliefs and evaluate evidence and arguments. For a broader, less intellectualist analysis of open-mindedness more in line with my use of the term here, see Song (2017, 65–84). https://doi.org/10.1080/00455091.2017.1335566.

7 Calhoun notes, with Millgram and Lars Svendsen, that meaning in life does not prevent boredom (2018, Ch. 6). The pursuit of projects experienced as meaningful, like completing an education or writing a book, can even be a source of boredom.

5

Shaping Behavior: Self-Control, Cultivation, and Emulation

Now it is time to bring the different strands of thought together. We shall consider the active aspect of change: How it is possible to change one's behavior, making it more pro-environmental, without impeding wellbeing? This requires us to look at the relevant behavioral science. But I shall also draw inspiration from philosophical ideas about self-cultivation. I will outline a psychologically realistic notion of personality formation or *Bildung* (as it is called in German), which accords with the classical versions of the idea, but can be made more precise by applying concepts and ideas from recent analytic philosophy of personal development and change.

5.1 The Bad News: How Not to Change Behavior

It has been a premise of this book that neither knowledge nor morality suffices for bringing about the changes in behavior required for green transition. Let us now look more closely at the evidence for this, not only to substantiate the claim, but also to consider possible qualifications. For it is highly implausible that knowledge and morality can make no difference whatsoever. Moral considerations seem to have played a significant role in large societal transitions like the abolishment of slavery or the enforcement of women's rights. Morality is far from powerless (for impressive examples, see especially (Appiah, 2010)). Knowledge has been said to simply *be* power; somewhat exaggeratedly, for sure, but knowledge can give one a very direct control over one's situation. Moreover, even though the version of the self-cultivation idea I will defend is intended to be more psychologically realistic, less idealized and moralized than rival versions, it is hard to imagine a genuine notion of self-cultivation that does not give some emphasis to the acquisition and use of knowledge as well as to morality.

First, however, let us consider the negative findings. In their seminal, but still not sufficiently widely acknowledged study entitled "Mind the Gap," Kollmuss and Agyeman amass evidence that rationalist models of behavior change are grossly inaccurate. Neither the possession of environmental knowledge nor environmental awareness leads to more pro-environmental behavior (Kollmuss & Agyeman, 2002). People with high levels of environmental awareness are not particularly willing to make lifestyle sacrifices (Diekmann & Preisendörfer, 1998). There is gap between attitudes and actions. Kollmuss and Agyeman convincingly argue that this gap cannot be explained by any simple model. Pro-environmental behavior is shaped—or hampered—by a very large and diverse set of (not easily delineable) factors—demographic, cultural, institutional, motivational, etc. Kollmuss and Agyeman find evidence that primary motives, such as altruistic and social values, tend to be "covered up" by more immediate, selective motives evolving around a person's own needs, for example, for being comfortable or saving money and time. This strongly indicates that to support pro-environmental behavior, we must tend to people's subjective wellbeing and make such behavior compatible with their need for simplicity, convenience, comfort, and efficacy. There is a directly proportional relationship between the cost (not just in money, but also in time and effort) and the willingness to act pro-environmentally (Diekmann & Preisendörfer, 1998).

Among other factors that might partially explain the attitude-action gap, Kollmuss and Agyeman point to the role of responsibility. This could be seen as a vindication of the widespread appeal to morality in discussions of climate change policy. But it is not the presence of moral beliefs or convictions or feelings of guilt that drives pro-environmental behavior. The point is more negative and more concretely psychological. Not seeing oneself as having any real influence—lacking an internal "locus of control"—reduces motivation and ability to act, and this partially explains the gap. People do not feel *concretely* responsible because they do not consider themselves able to make any positive difference. And when responsibility does function as a positively motivating factor, it is the responsibility one feels for one's own wellbeing and that of one's family, rather than the sense of a more general moral obligation (Kollmuss & Agyeman, 2002).

This analysis of the attitude-action gap confirms basic theoretical assumptions. It matches the emphasis on felt competence and autonomy made by the self-determination theory of Ryan and Deci (see Section 3.6). It further accords with the likewise influential theory of *self-efficacy* developed by Bandura: a person's belief in her capacity to exercise control over her own activity, based on her mastery experiences, observations of the successful actions of others, encouragement, and emotional arousal, is crucial to both motivation and her actual ability to carry out tasks successfully (Redmond, 2010). These theories also provide clues to what is needed for bringing about behavior change. At first, however, they help to explain the negative findings. What is lacking is not knowledge of the fact that individual action matters, still less abstract arguments that despite impressions to the contrary, individual can make a difference in a way that makes them morally obliged to try to do so (see Section 6.8). What insufficiently motivated individuals lack is first and foremost experience of their own successful agency.[1]

5.2 The (Somewhat) Better News: How We *Might* Change Behavior

Now let us take stock of what behavioral science tells about *efficient* means for bringing about behavioral change. This is important for mounting an adequate response to climate change. But it should be noted from the outset that it is still far from sufficient to know mechanisms and strategies that have been shown to work. Apart from questions of their existential and ethical acceptability, which will concern us later in this and the next chapter, it turns out that "what works" in many cases only does so in certain conditions. Making sure that these conditions obtain—and translating the general story about successful behavior change into actual, feasible steps—becomes a further difficult task. Behavioral science must also mind the gap between theory and successful practice. But before getting back to the more pessimistic assessments, let us review the positive findings, focusing on the research presented by Wood (2019).

Those who seem good at controlling themselves don't really do so—at least not in the way this is usually understood. They don't fight and resist temptation; they are less tempted in the first place. This does not mean that they are gifted with immunity or insensitivity to temptation. They are good at *stimulus control*—removing themselves from situations in which temptations are prone to arise or removing the opportunity to do a tempting thing (Wood, 2019, 69). This underscores the importance of *shaping the environment*, both when it comes to individual and collective strategies for behavior change. People who score high in self-control tests are also proficient at *automating*. They break and form habits through repetition, for example, going to bed early or eating healthy food stable behavior patterns that perpetuate themselves and do not demand any exercise of willpower (Wood, 2019, 75).

Repetition turns out to be one of three bases of habit formation. That behavior change requires arduous repetition is probably a major reason why it is so difficult. Often, a new and desirable behavior needs to be repeated so many times that the initial enthusiasm will have waned long before the new habit has been stably formed. The good news is that occasional gaps in the practice do not suffice to erase an emerging habit (Wood, 2019, 104). This is important, because the almost inevitable failures to comply perfectly with a new behavioral policy often bring people to lose confidence in their ability to change the behavior in question, and this may feel like a complete failure and cause them to give up. If one matters to get back on the track, the efforts spent already will not be wasted and the lapse into old habits remains temporary and insignificant. Repetition moreover changes the way an activity is experienced, making it feel easier (Wood, 2019, 111). This establishes a connection with wellbeing: we like to do what we are used to do, and we form desires to keep doing it or even doing it more frequently (Wood, 2019, 112).

The second base of habit formation is still more closely related to wellbeing. Habit formation is dependent on *reward*. To support the formation of new habits, the reward should, moreover, be *unexpected* (Wood, 2019, 116); it should stand out from the usual positive experiences one is likely to have. This matches the findings from psychological happiness studies emphasizing the role of novelty and surprise (see Section 4.4). The most effective habit-building rewards tend to be *intrinsic* to a behavior (Wood, 2019, 119). It is not so much the thought about the money you will be able to save or the endurance you will be able to build up that can make you carry on with living frugally or doing sport exercises. It is more efficient if what you do feels good or is closely associated with some immediate reward. For example, experiencing gratitude or appreciation from others can make one's actions feel pleasant and less arduous.

Third, Wood points to *context* as the third base of habit formation. Context generates forces on our behavior by both pushing or driving us towards doing certain things, as junk food advertisements may influence our eating habits. Sheer proximity of an object is a powerful driving force. We easily, sometimes almost inevitably, become used to grabbing what is easily within our sight or reach. But context also exercises force by restraining or impeding our actions—by creating *friction*, so to speak. Friction can impede the formation of good habits. It is a main barrier to pro-environmental behavior—the options are there, we know them, we know that we should use them, but they are more distant than options for less pro-environmental behavior or require more effort on our part. But creating friction can be used to build new and better habits, for example, by putting less healthy food slightly out of reach (Wood, 2019, 92).

One might ask—as Wood does herself—if it can really be true that self-control does not matter for building and maintaining new habits. For one thing, it seems quite commonsensical that the best way to improve one's behavior is to take control of one's life, setting goals and sticking to them, mustering willpower to resist temptations to do otherwise. This is such a fundamental part of received wisdom and folk psychology that it could hardly be all just a myth, one might think. Besides, there is a considerable body of psychological research that emphasizes the importance of self-control and posits a capacity for willpower, which it claims to be nothing less than "the greatest human strength" (Baumeister & Tierney, 2011). This line of psychological research has been applied to pro-environmental behavior change, with promising albeit not decisive results (Nielsen & Hoffman, 2021; Strömbäck et al., 2023).

Now, what are we to believe? Does self-control work or does it not? Self-control theory has come under quite severe attack as part of the so-called replication crisis in psychology (Carter & McCullough, 2014). I shall not, however, take any stance in this discussion. Baumeister may be right that the criticism has been exaggerated (2019). More important is how the findings, assuming that they can be taken as such, should be interpreted. The same goes for the findings reported by Wood. Conceptual clarification and philosophical analysis can contribute to resolve the apparent conflict between the different strands of research and help to bring out their practical implications more clearly. For the question is very much what is meant by "self-control" in the first place. Some psychologists have spoken, quite fittingly, of a conceptual crisis which may be more fundamental than the alleged replication crisis (Lurquin & Miyake, 2017).

To begin with, the view of habit formation espoused by Wood does leave room for a considerable amount of self-control, if this is understood a little more broadly. She recommends working deliberately with the three bases of habit formation. Refraining from going to the shopping mall or buying soft drinks that you would likely be tempted to consume arguably also takes some determination and willpower. The same goes for initiating a sequence of repetitive behavior. We can do it in the sense of *committing* ourselves to doing so (see Section 4.3). To this, a skeptic may reply that even commitment may not be powerful enough.[2] People make solemn resolutions to start doing something and tolerate the inconvenience it will bring but fail to carry on with actually doing it. But we may retort that repetitive behavior, as emphasized by Wood, need not be flawless. Commitment may not suffice to resist every temptation but still help getting one back on the track. Wood herself describes the repetitive element in habit formation as based on a willful decision and requiring that one put up a struggle:

> Before [habit kicks in], it's going to be some work. Until we have laid down a habit in neural networks and memory systems, we must willfully decide to repeat a new action again and again, even when it's a struggle. At some point, it becomes second nature, and we can sit back and let autopilot drive.
>
> (Wood, 2019, 102)

In short, although Wood insists, rightly, that exercising the desired behavior does not directly depend on willpower, she does imply that engaging in and carrying out effective behavior formation do require some sort of willpower or commitment and the ability to persist in the face of adversity. Self-control, on this picture, is a second-order rather than a first-order capacity: a capacity not for doing the right thing in the face of temptation, but for keeping on trying to become better at it.

Similarly, turning to self-control theory, Wood notes that even its originator, Baumeister, admits that "self-control may in general operate more by forming and breaking habits. It is thus mainly by establishing and maintaining stable patterns of behavior rather than by performing single acts of self-denial that self-control may be most effective" (De Ridder et al., 2012, 91). This is basically the same point, the only difference being that Baumeister and his colleagues call it self-control, whereas Wood reserves this term for single or ground-floor acts of resisting temptations.

In line with this broader understanding of the concept, recent self-control research has focused on the so-called "effortless" self-control strategies, like proactive avoidance of distractive environments (Gillebaart & De Ridder, 2015). The term "effortless" indicates that such strategies reduce the need for effortful inhibition of impulsive tendencies. It should not be taken to mean that the implementation of the strategies themselves requires no effort. Gillebaart (2018, 3) defines "self-control" as "everything that one *does* to steer one's behavior toward the desired end state." This clearly encompasses all the activities that Wood recommends. Moreover, for all the talk of "top-down" processes in the self-control theory literature, self-control in this sense does not require any acts of reflection, conscious decision, or even monitoring. It might suffice that one just *does it.*

I will get back to this when I consider the potential of personality formation, since it seems that classical theories of *Bildung* have been making the same point, recommending learning by doing and habituation through repetition. Yet although the power of such strategies and the inefficacy of knowledge and attitudes deserve to be emphasized, this does not mean that knowledge and attitudes cannot also be determinants of pro-environmental behavior change. Just like self-control may work in more indirect ways, knowledge and attitudes may likewise have an indirect impact on one's behavior and complement or enhance the habituation-centered strategies for change.

For a start, it should be noticed that Wood does assume that knowledge about the mechanisms and bases of behavior can make a difference. This is the knowledge she is keen on sharing, not just to enlighten fellow academics or policymakers, but also because it can be used by ordinary people to change their individual everyday behaviors. Knowing that it takes repetition, that lapses into old routines are no reason for giving up, and that one should choose or rearrange contexts to create less friction, makes behavior change more likely to succeed. Here, again, an otherwise inefficient means for bringing about behavior change turns out to be more potent considered as a higher-order capacity for selecting, initiating, and monitoring lower-order processes. Knowledge of behavioral science thus seems more immediately useful for behavior change than knowledge of wellbeing theory, which is, presumably, only relevant on the still higher-order level of behavioral design and policymaking (see Section 1.3). However, there may be a closer parallel, as for example knowledge of adaptation and coping strategies can also be used on a micro-level and even for individual self-improvement.

Knowledge—and, more generally, cognition—can also inform the emotion-driven processes that are largely responsible for achieving wellbeing and enabling behavior change. For example, it has been shown that *imagined stress*—feeling stressful emotions about our own potential future wellbeing—mediates the effect of relevance and temporal location of an event on our valuations of it. People are particularly concerned about their own future potentially stressful situations, whereas they tend to judge more neutrally (less temporally biased) from a third-person perspective (Caruso et al., 2008). This makes sense because our emotions, which presumably play a large role in these kinds of judgments, exhibit a strong temporal asymmetry, as we care more about what we may come to experience than what we have already gone through (Sullivan, 2018). It underlines the importance of providing vivid information about the likelihood of coming to live a satisfactory life with climate change mitigation (or, in case one wishes rather to use fear as a motivational factor, about the risk of coming to live less well if global warming is not curbed).

A further noticeable interplay between knowledge, emotions, wellbeing, and actions concerns the pleasure of doing, or trying to do, good. Knowledge of the harmful consequences of climate change can make one's efforts to combat it feel particularly rewarding. Acting green has been reported to elicit a "warm glow" (Taufik et al., 2015), and this functions as an immediate reward that can help habituate the actions in question. This further confirms the popular assumption that *participation* is important to pro-environmental behavior change (European Environmental Agency, 2023). Participation may likewise help finding creative solutions to sustainability challenges. It has an epistemic function apart from being a political right. But the role of participation in making green behavior feel pleasant and so self-enhancing should not be underestimated.

5.3 Limits of Behavioral Design

So much for the relatively good news. Unfortunately, several factors set limits to the efficiency and applicability of the principles of behavioral science in practice. Some of the most severe limitations are entailed, if not explicitly described, by the principles themselves. Although they are far from trivial in the sense of only stating the obvious—that if you want to change your behavior, you need to do whatever it takes—the factors they identify as decisive are not easy to control or bring about. Take the role of *context.* Behavioral psychology emphasizes the importance of an immediately supportive environment. But although we can directly redesign our environment to a certain extent, much of it is outside our direct control or otherwise not easy to change. More thorough redesign of our physical and social environment is a large and strenuous task that may not be easier to accomplish than the behavior change it is intended to bring about. Hence if you—and, importantly, not just you, but others around you—are not already able to concentrate on and persevere with making pro-environmental changes, the necessary changes in the environment are likely not going to happen. For all our knowledge about how we *could* change our behavior, we will remain stuck with it.

When this situation is described most pessimistically, it looks like an instance of a fundamental paradox or conundrum most famously found in Plato's *Republic*. Here, Plato lets Socrates argue that to make people act righteously, and create a good social environment, they, or at least those who are to be in charge, must be educated and formed in the right way. Yet such education and personality formation are only possible if the right social conditions already obtain. Only good cities can breed good people, but it takes good people to create good cities (Plato, 1992, 488aff).

Fortunately, this way of putting it is too pessimistic. We may, even if we are grossly imperfect people in far from optimal conditions, have some handle on the situation and be able to set a process in motion that will improve both ourselves and the environment. This is captured by the notion of *Bildung* (see Section 5.4). But the paradoxical description is instructive, since it illustrates if not the impossibility, then at least considerable difficulty of putting the seemingly powerful lessons of behavioral science to work. It is not that behavioral science is trivial and only teaches that if we want to change behavior, we ought to do whatever it takes. But it sometimes comes close, as Wood implicitly notices when she remarks "that self-control is simple when you understand that it involves putting yourself in the right situations to develop the right habits" (Wood, 2019, 76).

The practical limitations of behavioral science have been documented most strikingly in the case of behavioral design, more specifically the use of "choice architecture" or nudging to influence people's behavior. This has seemed to be a very promising strategy because it does not require that people themselves embark on a process of habituation or actively redesign their environment. It reduces the need for such activities and so for willpower and perseverance, by ensuring that the context is already supportive of the desired behavior, for example, by making sustainable food more visible or into the default option, like the dish of the day.

After a decade of strong enthusiasm, nudging now has a more mixed reputation. We shall consider the question of its ethical acceptability it the next chapter. The question here is how well it works. Metastudies indicate that little more than half of the nudging interventions that have been tried have had a significant effect, and that the effect is often not very large (Hummel & Maedche, 2019). It has even been claimed that there is no positive evidence for the effectiveness of nudging, if one also takes the publication bias—the tendency to report mainly significant and positive findings—into account (Mertens et al., 2022). There are many reports of specific kinds of nudges or nudges in specific domains that seem to have been more effective, and this is also documented in more specialized metastudies (e.g., Pandey et al., 2023). But doubts about the quality and replicability of the studies remain. In any case, it is now widely acknowledged that the effect of nudging and other behavioral change interventions depends on context as well as personal variables (Ingendahl et al., 2021). In line with what we noticed above, it also seems that attitudes, while being ineffective in themselves, can mediate and to some extent cancel the effects of nudges. And the effect of nudges that can change a very specific behavior may not spill over to other similar behaviors (Van Rookhuijzen et al., 2021).

The news about the limited efficiency of behavioral design is sobering, but no reason to despair. Again, the limitations seem to have been predicted by behavioral science itself. For one of its main lessons, it should be remembered, is precisely that context is crucially important. Echoing the Platonic paradox mentioned above—we need the right environment in order to improve the environment—behavioral science can remind us that context is not only a tool that can be put to use, inasmuch as we can control it, but also a factor that conditions our very attempt to do so.

Moreover, conceptual clarification is once again called for. Very different kinds of intervention are presented as behavioral design or choice architecture. For example, a much-reported study of attempts to reduce rates of commuting that all proved unsuccessful targeted interventions like sending letters about carpool registration and peer testimonials, a free bus trial, information on lost opportunities for savings, and the provision of personalized travel plans that provided customized information about carpooling and public transit options (Kristal & Whillans, 2020). Yet none of this is the kind of intervention we should expect to work well if we take behavioral science seriously; the interventions are hardly nudges at all. They are mostly attempts to change behavior through information; and the free bus trial is not something that leads to habituation or makes taking the bus easier or more affordable in the long term. Similarly, when the results are broken down into more specific types of intervention, the metastudy of Hummel and Maedche (2019) shows that defaults are often quite effective, whereas reminders are less so, and precommitment and implementation intentions and pre-commitment have mostly very little effect. Again, this is what should be expected. The problem seems to be not so much that behavioral science cannot work but rather that it is not properly understood or put to work properly.

It is thus also telling that the research-based literature on how to carry out behavior design or otherwise change behavior is preoccupied with methods for analyzing and structuring the task in question, testing and adjusting possible interventions, etc. (Hagger et al., 2020; Wendel, 2020). That is, despite talk of "*the* method of" behavior design, there does not seem to be any one single method for bringing about behavior change across contexts and types of behavior. The methods recommended are, again, higher-order methods aimed at selecting the appropriate method for the task at hand.

The context-dependence highlighted by the metastudies and replication failures of nudging experiments is closely paralleled by studies in moral psychology. Based on empirical research, philosophers like John Doris have mounted what has come to be known as the *situationist challenge* to virtue ethics (Doris, 2002). Virtue ethics is wedded to the assumption that good or appropriate acts flow from people's character, conceived as traits or dispositions that remain consistent across situations. A courageous person can be expected to act courageously in both war and peace, at home and abroad. Indeed, the very rationale for promoting virtues seems to be that they are globally or at least widely useful moral and social tools. Remember that Jameson views the development of green virtues—"nourishing and cultivating particular character traits"—as they are key to combating climate change (2014, 186).

However, the so-called situationists like Doris cite an impressive body of research suggesting that even small changes in the situation can influence the way a person acts, seemingly thwarting the impact of character traits, and so putting their very existence into doubt. One line of reply is to admit that virtue is rare and hard to come by, but still a meaningful ideal we might strive for. But while this may suffice to maintain the ethical significance of the concept of virtues, it renders it virtually irrelevant to the project of actual, short-term improvement of human behavior, and so to responding to climate change.

The story doesn't end here, however, and I have already hinted at reasons for being more optimistic about the possibility of systematically improving our own behavior and that of others. Let us now turn to the idea of self-cultivation. Since it likewise assumes that the development of personality traits and attitudes can lead to new and better actions, it might seem just as vulnerable to criticism from empirical psychology. Yet I will argue that rightly understood, it agrees with and can even be said to have anticipated current behavioral science.

5.4 *Bildung* to the Rescue?

The notion of *Bildung* has been adopted for many different purposes and interpreted in very different ways. Here, I take it to refer to a set of ideas—almost amounting to a theory—about the goal, conditions, and methods of human personal development, which was espoused by leading German writers and thinkers of the late enlightenment and early romantic period, notably Wilhelm von Humboldt, Schiller, and Goethe. These ideas are part of a larger tradition going back to the ancient Greeks and including figures from the British enlightenment, like the Earl of Shaftesbury (Horlacher, 2015). As we have seen, the notion was taken up, but also subjected to criticism, by Kierkegaard (see Section 4.4 and Klausen, 2018c).

Bildung is at root a *prudential* concept. The main rationale for propagating self-cultivation is that it is assumed to be necessary to live a satisfactory life, especially under conditions of emerging modernity. It is *instrumentally* necessary, because living well requires being able to cope with different situations and master the norms and habits typical of different social settings and groups from which one would otherwise remain excluded. It is moreover instrumentally *beneficial* in that it enables one to get the best out of what life offers, making one receptive to the values of art, science, and human interaction, and prepares one to actively seek and cultivate interests and enjoyments. But it is also considered *intrinsically* valuable, as the very activity of forming oneself is assumed to be inherently pleasant. Humboldt maintained that the development and exercise of all human powers, which he took to be the essence of *Bildung*, was also the source of the most profound happiness:

> It is in the pursuit of a single goal, and in achieving it through the use of all his moral and physical energies to its achievement, that the true happiness of a sprightly, powerful person consists [...] Pleasure is greatest in those moments in which man feels his individuality and creative energy at their highest pitch.
> (Humboldt, 1967, 14; 48; author's translation)

Remember how Kierkegaard's Judge Wilhelm describes how he takes comfort from persistently working on becoming himself. He enjoys striving to be a good civil servant and a good family man, taking aesthetic pleasure in forming himself as an individual as well as a human being in general.

In later times, *Bildung* has come to be understood more intellectually. It is sometimes equated with obtaining a particular body of canonical knowledge. In contemporary theorizing about the aims of education, *Bildung* is often related to critical reflection, responsibility, and political engagement (Klafki, 2000; Vásquez-Levy, 2002). Yet the classical notion of *Bildung* is broader. It calls for a balanced, harmonious development of all human powers and gives priority to action, emphasizing that morality and prudence must be learned first and foremost by *doing*. In current discussions, especially in educational science, it tends to be forgotten that *Bildung* was promoted as a reaction to what was perceived as the exaggerated rationalism of the enlightenment and the political extremism of the French revolution. Although they maintained the importance of achieving autonomy and living in accordance with reason, the proponents of *Bildung* wanted to rehabilitate the sensuous and emotional aspects of human life. They combined their belief in improving humanity with a realistic acknowledgment of the psychological and social constraints on such development.

Kant had insisted on sharply distinguishing morality from acting out of self-interest.[3] The proponents of *Bildung*, most notably Schiller, reacted by trying to reconcile duty and inclination.

Schiller objected to Kant that it is both possible and preferable to do good things while feeling good about them and to do them also in part because they *feel* good. Hence the goal of *Bildung* is to make otherwise exacting tasks feel easy by cultivating one's personality (Schiller, 2005, 149; 152). Very much in line with contemporary behavioral science, Schiller thinks that it will only be possible to turn doing the right thing into a lasting habit if it can also be experienced as enjoyable and aesthetically satisfying (Schiller, 2005, 150).

As noted above, the notion of self-cultivation can be traced back to the ancient Greeks.[4] In his *Ethics*, Aristotle anticipated both the ideas of *Bildung* and the lessons of behavioral science:

> For example: if you hold back from physical pleasures, and enjoy doing exactly that, you're moderate. [...] If you face frightening things and enjoy it, or at least aren't distressed by it, you're brave. If it distresses you, you're a coward. The fact is, being a morally good person is all about pleasures and pains. After all, we do bad things because they give us pleasure, and [typically] fail to do honourable things because they're painful. That's why it's important for us to have been brought up a certain way right from childhood ... so that we enjoy the things we should and feel pain at the things we should. That's what a good upbringing is. [...] So here's our basic claim: being a good person is a matter of [feeling] pleasures and pains in such a way that you do what's best; and being a bad person, the opposite.
>
> (Aristotle, 2020, 31–2)

The notion of *Bildung* likewise assumes that moral action must become aligned with, and exploit, human inclinations. It goes beyond Aristotle in not taking being a good person to be *merely* a matter of feeling pleasures and pains in conformity with moral norms. Schiller agreed with Kant that humans could and should rise above nature and be able to resist temptations (Gauthier, 1997). He and other proponents of *Bildung* thought that it should enable the individual to achieve a considerable degree of independence and self-control. Whereas Aristotle conceives of moral development as a mostly one-way process of socialization, getting in sync with the particular norms and circumstances of one's community, *Bildung* aims at developing higher-order mental and practical capacities that enable a person to adapt to a wider range of conditions, or, if need be, to rearrange them. Still, the differences should not be exaggerated. Aristotle and Schiller agree that it is not possible to resist temptation simply through reflecting on one's duties. Resisting temptation is also a skill and so must be acquired through habituation and the experience of pleasure. Moreover, while Aristotle appears more naturalist and Schiller more rationalist, the difference becomes less pronounced when one considers that Schiller follows Kant in identifying nature with the passive element in reality, while Aristotle views the distinctive actions of human beings and their exercise of reason as natural processes.

The core ideas of *Bildung*—expressed mainly in the texts of Humboldt, Schiller, and Goethe, including the latter two's fictional works (like Goethe's philosophically influential *Wilhelm Meister's Lehrjahre,* the epitome of a *Bildungsroman*)—can be stated as follows:

i. A human being can and must develop a self that is unique, coherent, and relatively independent (autonomous).
ii. Achieving selfhood and autonomy requires continuous effort, self-criticism, and self-restraint.
iii. Achieving autonomy is also a process of socialization. The materials for developing the self must be taken from the social and cultural traditions available as well as from encounters with the natural environment.
iv. The materials for self-development must, however, be individually appropriated and modified. *Bildung* is a process of mutual adaptation and attunement between the individual and its environment. While the materials must, to some extent, be taken from the social and cultural setting in which the individual is embedded, they should also include elements of other environments and traditions.
v. Though *Bildung* requires effort and experiences of challenges and limitations, it is driven by a concern for wellbeing (one's own, as well as that of others) and is a process which is itself experienced as fundamentally

pleasurable, rewarding, and meaningful (that is, intrinsically valuable and motivated).

As these statements show, even the classic, in some respects more realistic and practically oriented, notion of *Bildung* is still quite ambitious. When applying it to the task of responding to climate change, we should allow ourselves to abstract from its loftier commitments to special views about the overall goals of humanity and human existence.

Some might balk at (i), either finding that the very notion of a unique and autonomous self is problematic or, more specifically, that the ideal of *coherence* of the self is arbitrary (echoing my discussion of the value of a stable personal identity in Section 4.1). To this it can be replied that the independence and uniqueness attributed to the self in the *Bildung* tradition are only relative. It takes the self to be shaped in large parts by social and cultural factors and its relationships with others. For this reason, the popular English translation of *Bildung* as "self-cultivation" is misleading. It refers rather to a process of shaping and being shaped, best described as "cultivation" or "formation" simpliciter.

On the other hand, *Bildung* does remain committed to more than a deflationary notion of self and individuality. But this is probably also metaphysically defensible. Recent philosophy has gone a long way toward rehabilitating subjectivity (for just a few samples of this trend, see Bayne, 2010; Farkas, 2008; Klausen, 2020b; Searle, 1992). More importantly, regardless of whether it reflects any deep reality, it closely matches the way we continue to speak and think of ourselves and other people. We just are strongly and almost permanently concerned with ourselves and other individuals (as such). Hence it adds to the realism and practical usefulness of the *Bildung* approach that it conforms well to folk psychology and social practices as they currently are, without viewing the latter as unchangeable.

One of the problematic implications of some of the more radical metaphysical ideas put forward in response to the climate crisis is that they require us to abandon the anthropocentric viewpoint. In some cases, they even recommend us to put human beings on par with inanimate artifacts or natural objects (Wu, 2020). Yet it is very difficult to get people to effectively revise their fundamental conceptual schemes. And even if it is true that notions of subjectivity, selfhood, and individuality are just a modern Western invention and not central to the thinking of, say many African and Asian cultures, it is agreed on all hands that they are central to the thinking of the people I am mostly concerned with in this book: affluent Westerners. Moreover, for good or for bad, individualism has spread and is still spreading rapidly across the globe. Reversing this development is a very tall order. It would also a reignite a worry which I tried to defuse in Chapter 4, viz. that responding to climate change could put one's very personal identity and self-understanding in peril. *Bildung*, by contrast, offers a way to mediate between our inextinguishable tendency to care for ourselves and the need to care for the external environment.

Moreover, we should understand the importance *Bildung* attaches to building a coherent self as instrumental to promoting wellbeing. It need not be ascribed any intrinsic value, and it should be balanced against other concerns. But under normal circumstances, and probably still more under future circumstances of climate change, there are huge and obvious advantages to having a coherent self. Apart from making one able to function more effectively and withstand external pressures, it relieves one of the troubles that come, almost inevitably, from misalignment of the self. Tensions in the self are a notorious source of discomfort, both psychologically and existentially. This does not mean that we should strive at building a smoothly unified, tensionless self. Other personal qualities, like broadmindedness, many-sidedness, personality flexibility, and adaptive readiness, may be so important as to justify lessening the concern for coherence. *Bildung* aims at building a rich, complex, and nuanced self, which incorporates different and even seemingly opposed values and beliefs. It views personality building in analogue to musical composition, which is guided by the principle of variety in unity, working with harmonic tension, dynamics, and surprise effects to achieve something that is still a coherent whole. As with music, the ideal of building a coherent yet complex and varied self can be realized in very different ways, including more modern ones that allows for quite many tensions and dissonances within the self.

In its fullest sense, *Bildung* has often been described as a process of *becoming oneself*. This should not be dismissed as overly ambitious or practically irrelevant. As discussed in Chapter 4, there is worry that significant life changes, both willed and forced, can alter one's personality to such an extent that one becomes alienated from oneself. *Bildung* meets this worry, as it is a continuous and voluntary process of development, adaptation, and involvement that can be said to unfold the potential of the self. This is very similar to the notion of a *transformative activity* championed by Agnes Callard (see Section 4.1), also because it need not be the result of any clear prior intention or plan. The process is voluntary not because it depends on a particular decision but because it is accompanied by a sense of agency and personal satisfaction. By engaging in it, the person who undergoes *Bildung* proleptically identifies and sympathizes with her only vaguely anticipated future self and life.

5.5 Getting It Right: *Bildung* and the Situationist Challenge

The idea of *Bildung* is obviously attractive. But is it useful in the context of climate change? Does it not suffer from the same flaws which we have repeatedly found in other approaches? It is concerned with character building; yet we have seen that the circumstances matter more for than personal dispositions. And does it not describe an ideal state, rather than the actual means for approaching it?

These are reasonable worries. *Bildung* does, on the surface, appear quite similar to virtue ethics, which faces the situationist challenge and seem to presuppose an already-supportive environment. However, there are important differences. *Bildung* is not the idea of a one-sided character development aimed at building a one-size-fits-all set of global dispositions that will be valuable regardless of an individual's actual life circumstances. On the contrary, it is the idea of an interplay between character and circumstances, and a process of discovery: What are the obstacles and affordances to development and flourishing in one's environment? What are the possibilities for reshaping the environment or seeking new pastures? *Bildung* moreover explicitly requires favorable cultural and natural circumstances to succeed. Schiller was keenly aware of this, seemingly also having foreseen the limits of hedonic adaption (see Section 3.7):

> The germ of the beautiful [which Schiller views as intimately connected with *Bildung*] will find an equal difficulty in developing itself in countries where a severe nature forbids man to enjoy himself, and in those where a prodigal nature dispenses him from all effort; where the blunted senses experience no want, and where violent desire can never be satisfied.
>
> (Schiller, 1982, Letter XXVI)

"Personality formation" may be a more appropriate translation of *Bildung* than "self-cultivation" because it refers to a process that involves both one's own efforts *and* external influences. Indeed, it takes even one's own efforts to be prompted and shaped by natural and cultural factors. It does not assume the presence of a clear intention and strong willpower from the outset, but rather describes how these may emerge gradually, in the interplay between self and world.

This may suffice to accommodate the situationists' points, but at the cost of making *Bildung* less practically relevant. For it seems to make *Bildung* vulnerable to the same criticism as the behavioral science of habit formation (see Section 5.3). It works when the conditions are right, but our current inaction predicament indicates that the conditions are not right, that they are not sufficiently supportive, but rather cater to those of our inclinations that are neither pro-environmental nor conducive to lasting improvement of our personal capacities for change. We are situated in an environment saturated by commercial and economic growth interests, and most people appear weak-willed and incapable of significant self-improvement. They have adapted to the environment, but this has made them maladaptive to the current situation, in which radical change is needed.

My response to this criticism is similar to the one I gave on behalf of behavioral science of habit formation, albeit with some modifications. First, answering the situationist challenge by admitting that we are concerned with an ideal—an exceptional condition (Merritt, 2000, 367f.; compare Appiah, 2008)—rather than a common psychological reality may not be quite wrong, at least in the present context. It should be remembered that we are not concerned with ethics as such. In contrast to virtue ethics, *Bildung* is not supposed to be an alternative to utilitarianism or deontology. Now it may be thought that an idealized notion is even less helpful to the task of actual behavior change than it is to ethical theorizing and judgment. But it is agreed on all hands (also by Doris) that consideration of, and attempts to emulate, ideal behaviors can help us some way toward realizing them. Moreover, acknowledging one's own limitations, and those posed by nature and society more generally, is central to *Bildung*. Traditionally, this acknowledgment has been supposed to come about due to personal experience. But scientific evidence may also contribute to a realistic—*and* practically relevant—assessment of what one can achieve and how. Appiah imagines how a person striving to be virtuous might welcome the psychological research cited by the situationists:

> Reading about these experiments will only remind her that she will often be tempted to avoid doing what she ought to do. … Each time she sees someone who needs help when she's hurrying to a meeting, she'll remember the [findings showing that people who are in hurry are less likely to show compassions] and tell herself that, after all, she's not in *that* much of a hurry—the others can wait.
>
> (Appiah, 2008, 49)

But am I not now saying that knowledge drives behavior change, contradicting both myself and the impressive body of research that has established the existence of a gap between knowledge and action? While I am indeed saying that knowledge can *somehow* drive action, I am not questioning the knowledge-action gap; I am suggesting a way to bridge it.

The idea is not that individual possession or intellectual grasp of the ideal of *Bildung* (or, for that sake, knowledge of behavioral science) drives one toward its realization. This would be wishful thinking. It is rather that understanding the ideal of *Bildung* makes it possible to identify, motivate, and orchestrate more specific tasks and strategies that can contribute to its realization. As an *ideal*, it works on a quite high and general level: as a guideline for policymakers, educators, and behavioral designers, but also for individuals faced with the overall management of their life. Yet *Bildung* also denotes a less perfect, psychological, and sociological reality: it is already a fact, and has been so for a very long time, that human beings are formed in a process of mutual adaptation and adjustment, driven by a concern for wellbeing. The suggestion is that we use our knowledge of this process to control it and harness its power, directing it toward the kind of environment and behavior we would prefer. And *Bildung* does posit mechanisms and strategies for approaching the ideal, and thus for its own realization. These are very much the same that are recommended by contemporary behavioral science: emulating exemplary behavior (we will get back the role of social influence in the next section), repeating behaviors, and doing it perseveringly, making exemplary behavior rewarding, redesigning the environment as far as possible and eschewing it when it cannot be changed, cultivating awareness of supportive factors or making them more salient, etc.

Third, although *Bildung* does involve the formation of specific inclinations and habits, it is as such not a specific disposition or set of such dispositions. It is a higher-order capacity for selecting, initiating, monitoring, and controlling lower-order processes. *Bildung*, it should be remembered, was put forward as a response to the condition of emerging modernity: to help the individual come to terms with a world that is larger, more diverse, and unpredictable than the cosmos and *polis* for which Aristotle conceived his ethics. It can be argued that the situationist challenge does not apply so strongly to the original Aristotelean version of virtue ethics, because it assumes that people will only confront a limited range of situations (although this may seem to make it defective in another way—overly relativist and confined to a particular social setting).

What has been cast most strongly in doubt by empirical psychology is the existence of stable *mid-level* dispositions to behave in a certain, but still fairly general, way (e.g., acting courageously or generously) across the full range of situations in which the behavior would be appropriate (Merritt, 2000). It is not in dispute that people have stable *low-level* dispositions to behave in *particular ways* in *particular* circumstances. And it seems likely, and has not been challenged empirically, that people can have *high-level* dispositions and capacities to behave in certain general ways under virtually all circumstances, using particular styles of thinking or problem-solving, exhibiting more or less self-control, etc. For example, intelligence is considered a stable personality trait. More relevantly, it has been suggested that *adaptive readiness* or adaptability (Savickas, 2013), *self-esteem* (Hank & Baltes-Götz, 2019), and *openness* (DeYoung, 2015) are likewise relatively stable personality traits. A similar lesson has emerged from the debate on self-control: its proponents concede that it not very effective at performing single acts but may help instead establishing and maintaining patterns of behavior (see Section 5.3 above). This has obvious practical implications for green transition: We should, one the one hand, focus on specific tasks and behaviors, going for more or less ad hoc solutions, with no illusion of being able to upscale them widely, and taking this into account in our cost-benefit calculations. On the other hand, we should try to foster general dispositions to self-improvement, self-control, and adaptation readiness (for more on this as a practical policy, see Chapter 7).

Understanding *Bildung* as being concerned with the formation of general and higher-order capacities aligns with some of the most promising responses to the situationist challenge from virtue ethics. Kamtekar has argued that character traits should not be conceived as "independently functioning dispositions to behave in stereotypical ways, dispositions that are isolated from how people reason." The concept of character in virtue, she contends, is "holistic and inclusive of how we reason: it is a person's character as whole" (Kamtekar, 2004, 460). Christine Swanton's "target centered" version of virtue ethics (2021) focuses less on character but is compatible with idea of *Bildung* in suggesting that we should strive for whatever it takes to ensure an appropriate response to a situation, making this a matter of *overall* virtuousness rather the presence of a single virtue.

But what about the fundamental worry about the practical relevance of *Bildung*? Does it not assume that people are already sufficiently competent or cultivated—that is, positively disposed—to engage themselves in or be moved by the process of cultivation? If they are not, then the theory may be descriptively correct, but the "cultivation" we can expect will not be beneficial to green transition. It will consist in a further reproduction and refinement of consumer lifestyles, as people continue to become socialized into a commercial and capitalist culture and develop habits and preferences accordingly.

This is correct. *Bildung* does assume that people are fundamentally capable of positive change and self-improvement. It assumes that they possess some degree of autonomy and self-control. It also assumes that they are placed in circumstances that are not completely adverse and to which they are not completely maladapted. It provides a recipe for harnessing and exploiting existing opportunities, using available mental and social powers, not for pulling oneself up by the bootstraps. (Remember how Schiller emphasized that if the environment is not sufficiently friendly and the personal resources not sufficiently plentiful, *Bildung* cannot get going in the first place.)

This is not, however, a problem peculiar to *Bildung* or strategies for pro-environmental behavior change. It is a condition of normative theorizing and projects of improvement quite generally. For example, epistemology must assume that our most basic sources of belief formation are sufficiently reliable (Alston, 1993) and that we are placed in a sufficiently supportive environment (and not, say, subject to the deceit of a Cartesian evil demon). In fact, the situation we are facing in the field of behavior change is considerably more favorable. For here there is no problem of circularity or fundamental skeptical challenge. We can use the established practices of belief formation to determine whether and to what extent the requirements for *Bildung* can be met. What is at issue are empirical assumptions about human psychology and the current social environments. These are open to empirical study. And we have already seen, in Chapter 2, that the historical record shows that people can indeed change rapidly and fundamentally, and that the processes of self-enforcement and mutual adjustment described by *Bildung* are indeed possible. There are many examples of people either themselves creating new life circumstances, adapting to them and internalizing new norms as they go along, or of people being forced or pushed to do so.

Admittedly, there have also been many cases in which seemingly promising projects of improvement have floundered, sometimes spectacularly. And often these failures are most plausibly explained by a mismatch between the well-intended plans and the available resources, not least in terms of the people involved. So what if the conditions, and the nature of the transitions in question, are now so that change is more difficult to bring about? Green transition seems to cater less to human psychology and attract less support from the dominant social and economic powers than did the introduction of industrial and capitalist forms of production or the establishment of the welfare state in postwar Europe.

There is something to this worry. Green transition is undoubtedly more difficult than, say, the adoption of market economy in China has been since the rewards are less tangible and ubiquitous and the burdens more salient. This is the explanation I have given for its stalling, and the reason why I have urged realism. But we have little reason to think that the difficulties are insurmountable. In most cases of floundering or stalling processes of societal improvement, the failure to move forward can just as well be explained by a neglect of *Bildung*. This should not be understood as the lofty notion that if only people had become more cultivated and virtuous, positive change would certainly have ensued. I speak of *Bildung* in the practical and psychologically realistic sense outlined above. Some projects may have been outright infeasible. But others that floundered may have done so because too little was done to create a mutual adjustment of the actual tasks and actual resources, including people's readiness and ability to comply.

It is hardly an issue whether people are generally able to pursue a process self-cultivation of the sort required by green transition. They obviously are. We have seen that the changes they will have to undergo are in many ways relatively minor and superficial (Sections 2.5 and 2.6). The issue is rather if the current, specific constellation of tasks, powers, interests, and possibilities for reward and identification makes it likely that they can in fact do so. But this also seems quite likely. First, there are also examples of quick and significant behavior changes in cases in circumstances comparable to those we are considering here. The decline in smoking since the 1960s shows that it is possible to change a behavior that is supported by reward mechanisms, cultural norms, and economic interests.

There are of course important differences. Climate change does not pose a comparable health risk to the individual. Nevertheless, the reduction of tobacco use is a striking example that much can be achieved by altering social contexts and incentives (Cummings & Proctor, 2014). Information about health risks has no doubt played a substantial role; knowledge that engenders fear for one's personal wellbeing is a mainspring of human action. However, there are strong indications that peer influence and the general culture shift—smoking becoming less cool and smokers less exemplary role models—have been responsible for much of the decline in smoking (Frank, 2020). It has probably been necessary for it and might even have been sufficient.

Second, it is only partially true that the current circumstances are not conducive to pro-environmental behavior change. Kate Soper is probably right that the current consumerist lifestyle is already widely, albeit tacitly and subtly, perceived as unsatisfactory. Soper admits that the discontentment is seldom direct or pronounced. It is manifested first and foremost in a dissatisfaction with what can be seen, from a more detached stance, to be *consequences* of this lifestyle: overwork, stress and insecurity, long transportation times, a shortage of real variation and aesthetic affordances and the loss of landscapes, spaces, and forms of social life for which people feel nostalgia (Soper, 2020, 54). But it is also manifested in more direct yet implicit feelings of dissatisfaction with the current life form. These are present in even rich people, who may not experience existential anxieties or practical inconveniences, but instead feel ashamed and so are less able to really appreciate their life (Sherman, 2017).

Because even clear knowledge of the causal connection between consumerism and its unpleasant consequences may not suffice to shun the former, this manifestation of dissatisfaction may seem to be largely ineffectual. However, making the connection more salient may help. Building negative associations is known to be a quite powerful mechanism. Acknowledging the connection also gives policymakers reasons to assume that curbing consumption will not be so difficult. As for the rudimentary, implicit feelings of dissatisfaction with consumerist lifestyle, these can be boosted by cultivation of both a negative and positive experiences. Fostering awareness of the tacit displeasure with consumerism can strengthen it and make people poised for action. Fostering awareness of the pleasure that might come from less consumerist lifestyles can likewise ease the transition. In short, there is much in the current situation that *Bildung* can work with. But of course it must be put to work. There must be some kind of pro-environmental agency at some point in the process, which enables it to exploit affordances for positive change.

5.6 *Bildung* and Behavior Change as Social Process

Bildung is to a large extent a social process. It is autonomous, but autonomy is not an innate power; it is also gradually fostered in a process of socialization (see feature (iii) above). By emphasizing the importance of role models and social learning, the *Bildung* tradition revitalizes ideas expounded by Aristotle (and, still earlier, the Confucian tradition in Chinese philosophy[5]). The question we have been pondering is how the process of gradual improvement can be initiated and set off in the right direction. We assume that people become good by doing good things. But how can they know what the good things are before they have themselves become good? Especially, how can they come to know what is good before it feels good to them? The answer from the *Bildung* tradition is that they emulate the behavior of others. If those who are already inclined to do good, and act accordingly, also appear exemplary, and so worthy of emulation, other people will be disposed to imitate their actions, even if it causes them initial displeasure or practical inconvenience.

Again, it is important to keep in mind that it is not morality that does the work (and there is nothing in the mechanism described here that ensures that the resulting behavior will be good in the sense of being morally right). The people whose behavior is emulated function as role models not because there are morally exemplary, but because they *appear* worthy of emulation. They are perceived as admirable and enviable, often because of their social standing, sometimes because of their personal or even aesthetic qualities (self-confidence, charm, humor, apparent skills at mastering life, etc.). We have seen that vivacity, engagement, and integrity are also perceived as desirable personal characteristics (Section 4.4). Smoking is not a morally valuable activity; it is not even prudentially valuable. Yet the habit of smoking is also acquired in the way described by the *Bildung* approach: it is copied from someone who appears cool or successful (according to often idiosyncratic success criteria); the urge to become like this person, or gain her acceptance and access to her social group, is strong enough to carry on with the initially unpleasant activity until it eventually feels good. However, if the "magnetic" personal characteristics can be aligned with a genuinely desirable—for example, pro-environmental—behavior, the process becomes one of *Bildung* in the positive, normative sense, a process of improvement.

The power of social contagion has been described extensively by the economist Robert H. Frank, who strongly recommends that it be put to more extensive use as a tool of public policy, not least to foster pro-environmental behavior (Frank, 2020). Frank argues, much in the spirit of *Bildung*, that being inclined to emulate others is prudentially rational. Apart from the more immediate, experiential rewards it may bring, it helps us navigate our social environment, fostering social inclusion, deepening social ties, and supporting social identity. It also helps (again, very much in the spirit of *Bildung*) calling attention to new possibilities or hitherto unrecognized risks and limitations (Frank, 2020, 50).

The contagion metaphor is apt not just because behavioral dispositions spread from person to person, but because the spread is amplified within a group. The analogy to cases of infectious disease is quite striking, as behavioral contagion is similarly mediated by factors like the duration or intensity of the contagious behavior, transmission probability, and susceptibility (see Kurcharski (2020) for a general analysis of processes of contagion in different domains). If a certain behavior becomes slightly more prevalent in a peer group, this increases the likelihood that it will spread further significantly. Thus it makes sense to speak of *cascade* behavior.

The concept of a *tipping point* beyond which a hitherto rare behavior spreads very rapidly was introduced into mathematical sociology by Thomas Schelling (1971) and popularized by Malcom Gladwell (2002). Later attempts to debunk the idea (Watts, 2011) have focused mainly on Gladwell's idiosyncratic gloss on it, especially his notion that social changes depend on a very few exceptionally influential individuals. There is nothing wrong with the basic model: people have certain thresholds for adopting a new behavior, depending partly on the number of people in their environment who already exhibit it; small increases in this number, and/or small changes in this threshold, may unleash a rapid chain reaction (Frank, 2020, 69). Real-life examples of this dynamic can be found in cases of swift and unexpected regime changes and the rapid spread of the #MeToo campaign in 2017. While the latter also seems to vindicate parts of Gladwell's version of the tipping point idea—a single a tweet by actress Alyssa Milano is generally recognized as *the* tipping point that sparked the #MeToo movement (Brittain, 2024)—empirical studies indicate that peer effects are strongest among groups of close friends (Ali & Dwyer, 2009).

It is important to notice that there is no automaticity in processes of social contagion and no general rule for when and how fast a certain behavior spreads among a population. People have different thresholds and may be more or less susceptible to peer influences in different domains. This is why cultivation—*Bildung*—matters. Cultivating an awareness and appreciation of certain kinds of behavioral innovation may make people more prone to imitate pro-environmental behaviors once they become salient in their environment. Conversely, fostering a certain disinclination or resistance to simply go with the flow may make them partly immune to pressures to consume or otherwise act environmentally unfriendly.

There is, however, also some apparently bad news to be found in the field of behavioral contagion. Negative behaviors are more likely to produce contagions than are socially desirable behaviors. It is easier for people to choose the latter independent of other people's choices (Goss & Cook, 1996). However, it can be argued that pro-environmental behavior does not count as socially desirable in this sense. For though it may be verbally endorsed and officially recommended, it is still not a strong social norm. Environmentally unfriendly behavior is *de facto* still strongly socially supported and widely perceived as desirable. (This is why it could be argued that the overall social environment is not conducive to pro-environmental *Bildung* processes.) Hence at least before it becomes more widely adopted, it can still be seen as a kind of "rebel" behavior that both require and support social contagion.

Frank expresses his overall recommendation in a way that epitomizes the idea of realistic *Bildung*: "We could adopt simple incentives to choose in ways that would foster environments more likely to bring out the best in us" (2020, 57). This is quite similar to my response to the charges of practical ineffectiveness and powerlessness of choice and character: we can improve ourselves not by mustering willpower or deciding to act in a better way, but by deciding to employ a strategy that might eventually lead us there, fostering more supportive environments also by exercising peer pressure ourselves. Is should be remembered that *Bildung* is about mutual adjustment and development. One takes something from one's social environment but also gives something back, and so contributes to continually reshaping it.

The large role played by the social context may at first seem to be a reason for further skepticism about the prospects of fostering pro-environmental behavior through self-cultivation. It appears to reinforce the point that context matters more than character and to leave virtually no role for the self. But this is too pessimistic. Self-cultivation or *Bildung* is not a process that is independent of or goes against social trends. It exploits and thrives on them. Social support makes *self*-cultivation easier. If others push—or merely go—in the same direction, it becomes easier to act on one's own attitudes and beliefs. It can also bring one to realize them in the first place, that is, reveal one's real preferences to oneself. Sunstein describes how a change in the normative landscape can *unleash* people, making them feel free to disclose their experiences and act as they wish (2019, 11). *Bildung* theory (and the notions of aspiring and endeavoring (Section 4.1)) adds the important corollary that the preferences may be further shaped and refined in the process of being revealed; the person undergoing change may only incompletely know what she would really like, but gradually discover it in a process of social learning.

Self-control, understood as a personal power to withstand temptations, might seem like a myth. Yet given the right kind of social support, people are in fact able to control their behavior and realize their long-term objectives (Van Dellen et al., 2015; Nielsen & Bauer, 2019). This is why sports exercises are often best done in groups, or at least with a training partner or coach.

Moreover, the phenomenon of behavior cascades and tipping points makes it plausible that character can play a role, after all. The empirical studies that put into doubt the existence or power of character traits do show that some individuals—often around 10 percent—*are* able to act out of a stable character, that is, act consistently across contexts. Certain individuals and certain more salient pro-environmental behaviors may trigger a behavior cascade, especially if other individuals in the environment are sufficiently susceptible to peer influence. This may again depend partly on their values and attitudes (Liao et al., 2022), which are therefore not completely irrelevant.

5.7 Return of the Environmental Humanities?

It might now also seem that the view typical of the environmental humanities has been at least partly vindicated. Social contagion and the existence of tipping points make it less unlikely that a change in mindset or relation with nature could effect significant changes in actual practices, perhaps quite soon. The historical cases of slavery and women's rights give reason for optimism. Likewise, it would seem that my justification of the *Bildung* approach also applies to the environmental humanities I have dismissed as overly idealistic. Although fostering specific attitudes does not in itself support specific courses of action, it does support changes in values, attention patterns, and strategic decisions that can alter what is experienced as pleasant and unpleasant and lead to changes in the context—which might then, in turn, change one's behavioral dispositions. There may even be a place for posthumanist thinking. Coming to understand ourselves as merely parts of comprehensive reality rather than its center or dominating force may also make it easier to adopt sustainable patterns of behavior.

This is admittedly so. In addition, we should acknowledge further potential benefits of coming to think in less anthropocentric ways, which may for example benefit mental health. But it is at most a partial and qualified vindication of the thinking typical of the environmental humanities. First, it should be noted that the normative justification of seeking to cultivate a less anthropocentric mindset must eventually be anthropocentric: it might be good for humans and other sentient beings if humans come to think of the world in a less anthropocentric way, but it is only good insofar as this is actually the case. Second, it is important to keep in mind how posthumanist thinking might work, if it can work at all. Like all other attitudes and ideas, it needs to be translated into concrete behavioral dispositions. It must be supported by emotional reactions and social structures, and the behaviors which it calls for must be established by repetition and rewards. It also needs to be directed and calibrated to match actual needs and opportunities vis-à-vis climate change mitigation. As we shall see in Section 6.6, many actions that express respect for, or appreciation of, nature do nothing to reduce emissions, and vice versa. There may be no technological quick fixes, but there are many indispensable tools and strategies for mitigating climate change that "exploit" nature, as for example wind- and hydropower do. Overall, we need an active, interventionist, economic approach. It may be correct that even radical environmentalist thinking, inspired by, for example, ancient Chinese Daoism and its ideal of "non-doing" (*wuwei*), can accommodate a concern for "restorative interventions" (Nelson, 2020). But we must accept, and actively commit ourselves to, quite drastic interventions, if we are to stand a chance of tackling climate change.

Lastly, the critical-historical parts of the posthumanist narrative are of little practical use. Coming to learn that present exploitation of nature stems from, for example, theistic religion or Cartesian dualism contributes nothing to changing our present behavior.

Regardless of how much of the environmental humanities that can be vindicated, or reconciled with *Bildung*, the central point remains: We must work with the human factor. Self-cultivation is neither simple nor easy, and it could not work if the circumstances were hopeless. But in the present situation, it does hold a realistic promise to contribute to and boost the green transition.

Notes

1 Wood provides further evidence that what we know and what we do tend to remain disconnected (2019, 57).

2 It should be noted that research apparently showing commitment to be powerless has operationalized commitment as an *avowal* to do something (see, e.g., McCaul

& Kopp, 1982); in their taxonomy of determinants of pro-environmental behavior, Valkengoed et al. define commitment as "people pledging or promising that they will engage in a certain pro-environmental behavior (or abstain from an environmentally harmful behavior)" (2022, 1485). Yet such avowal or pledging is arguably neither necessary nor sufficient for genuine commitment, which requires a higher degree of internalization and identification (see Section 4.4).

3 Recent scholarship on Kant tends to interpret him as holding a more nuanced view. Correspondingly, some interpretations of Schiller have depicted his view as an attempt to modify, rather than as an alternative to, Kantian ethics (Guyer, 1993; Gauthier, 1997). However, in central places Kant does seem unambiguously committed to a stern distinction between acting from duty and acting from inclination, notably in the *Groundwork*: "To be beneficent where one can is a duty, and besides this there are some souls so sympathetically attuned that, even without any other motive of vanity or utility to self, take an inner gratification in spreading joy around them, and can take delight in the contentment of others insofar as it is their own work. But I assert that in such a case the action, however it may conform to duty and however amiable it is, nevertheless has no true moral worth, but is on the same footing as other inclinations, e.g., the inclination to honor, which, when it fortunately encounters something that in fact serves the common good and is in conformity with duty, and is thus worthy of honor, deserves praise and encouragement, but not esteem; for the maxim lacks moral content, namely of doing such actions not from inclination but from duty" (AK398). Likewise, Schiller does complain about the "harshness" of Kant's view [GD 206/545] and even fears his requirement of "subjection" and "submission" to reason can lead to a tyrannical reign (*AE* IV.2/19).

4 Although there is no historical connection, it should be noted that the idea of self-cultivation is still older, as it also figures prominently in ancient Chinese philosophy (see Weingarten, 2015).

5 See especially the *Xunzi* (Hutton, 2014).

6

Acceptable Transitions

Now we shall finally consider the moral constraints on pro-environmental behavior change. We shall look at both possible constraints on local interventions and moral concerns relevant to more general policies and their implementation. Apart from considering purely normative—that is, mostly moral—restrictions on what can be done to support the green transition, we shall also look to moral psychology and the *perceived* acceptability of strategies and interventions.

Climate ethics has focused strongly on individual obligations to act pro-environmentally or abstain from actions that might contribute to climate change. By contrast, I will consider whether and how far interventions from the outside—actions of *others* aimed at bringing about behavior change—can be justified. This covers both cases of government action, like laws, regulations, financial redistribution, and communication campaigns aimed at arousing fear (McQueen, 2021), as well as cases of more local or informal intervention, like peer pressure. It is, admittedly, to stretch the meaning of "intervention" to use it also to refer cases where someone urges a friend to change behavior or exerts influence on her simply by being a strong role model or a live illustration of how to live sustainably and well. But this is what I will do. My focus is on all kinds of actions that could contribute to making someone adopt a more pro-environmental behavior and which could be perceived as intrusive, paternalistic,[1] or otherwise disrespectful of personal boundaries or individual liberties.

We have already seen that tinkering with people's preference formation is considered by many to constitute a problematic violation of their autonomy. In public debate, more general worries have surfaced. Being free to choose and exercise everyday behaviors, including those related to consumption and movement, is widely considered essential to living a truly human life; having a *private sphere* is seen as a fundamental human good. Thomas Scanlon has argued that we have a right to privacy grounded in "the special interests that we have in being able to be free from certain kinds of intrusions" (Scanlon, 1975, 315).

The worry that behavioral design and other interventions aimed at bringing about pro-environmental behavioral vitiate personal autonomy is the predominant concern in discussions of acceptability. The main question is whether otherwise well-justified measures could have negative moral side-effects in terms of restricted autonomy that require them to be constrained or abandoned. We shall also, however, briefly consider the possibility that inventions can be unacceptable because they lack sufficient moral justification in the first place. This brings us closer to environmental ethics as such.

6.1 Only Soft Interventions?

One popular line of response draws a line between prohibitions, legal requirements, and economic incentives on the one hand and softer forms of social influence on the other, which maintain the right and possibility of the individual to choose to do as she wishes. The latter have been defended by proponents of nudging like Thaler and Sunstein, who have coined the phrase "liberal paternalism" (2008). This refers to a view that countenances "hard" interventions, like prohibitions, in some important cases but tries to uphold the liberal idea of protecting the private sphere and preserving individual freedom by only influencing everyday behavior in softer ways, albeit in ways that are highly likely to alter it on a group or population level. Some people may resist the nudges, but they are generally expected to work.

It makes good sense to distinguish prohibition and legal requirements from other means of promoting pro-environmental behavior, even though it is not clear that they differ fundamentally in terms of their normative justification or acceptability. However, placing financial incentives in the first category seems to reflect a libertarian prejudice that taxation is a particularly substantial intrusion into people's private lives. From a more neutral standpoint, it seems to belong rather in the second category of softer means of behavior promotion. Making it more costly to eat red meat or travel by air does not prohibit anyone from doing it. That it may nevertheless—in practice—force people to reduce or give up these behaviors is of course correct. But to care for this is to care for positive or substantial, rather than merely negative freedoms, and this goes against the libertarian view.

The distinction between hard and soft means of promotion is hardly sufficient, or sufficiently clear-cut, to resolve the acceptability issue. Some "hard" interventions may be at least as easily acceptable as some "soft" ones. An example could be mandatory carbon labeling or climate disclosure, which, in effect, function much like nudges, as they leave it up to the individual decide how to act (although altering behavior through information is usually distinguished from nudging). On the other hand, there is a tendency, also among environmentalists, to view nudging as *more* problematic (and in a way "harder") than altering behavior through laws and regulations. This stems from a perception that they are less transparent and more manipulative, and that by using nudges one does not treat people as rational, autonomous persons.[2] It can be okay, on such a view, to make it practically unaffordable for people to fly, if the decision to do so is democratically legitimated and the taxation rule applies to all. But it would be less okay to try to influence their actions by catering to unconscious biases or redesigning their environment behind their back.

Yet nudges and behavior design are as such neither softer nor harder than other kinds of intervention. It depends on how they are carried out. If one worries that they could be abused or come to be used excessively, one might demand that their use become regulated by a democratically enacted law. This would put them on par with other politically enforced regimes like taxes or driving rules. Like these, they would not, however, be individually approved. They could still be strongly resented by some of the affected persons, just like some find it unjust that they are legally required to pay a certain tax or not allowed to drive above a certain speed on the motorway.

In a rare attempt to formulate an ethics for behavior change interventions, Jun and colleagues have suggested that designers should be obliged to provide those who will be affected by an intervention "with well-balanced suitable information for rational and considered thinking so that they would consent to having their actions influenced in such ways" (Jun et al., 2018, 119). They moreover have suggested that the design process itself should involve those potentially affected, recommending the use of the so-called participatory design. There is much that speaks for such an approach. Involving "users" of the design may be epistemically beneficial since they will often contribute relevant knowledge of what may or may not work. Besides, as we have seen in Chapters 4 and 5, participation can strengthen identification and self-efficacy and enhance wellbeing, thus making it easier for interventions to catch on.

However, requiring that all behavior design should be participatory or that actual consent be obtained in all cases may not be realistic. It would make behavior design cumbersome and prevent it from being widely and rapidly deployed, which is what is needed for a speedy green transition. A more realistic requirement, which still captures the spirit of the proposal of Jun and colleagues, would merely demand that those affected *would* most likely consent if they became informed about the intervention or that they would certainly do so if they were fully informed and rational. That is, *hypothetical* consent should be enough. Most interventions will probably meet the latter requirement. We can assume that they will not impair the wellbeing of those affected. For they will probably work by making the desired behavior easier and more rewarding, and also support long-term wellbeing, for example, due to health benefits from adopting a more sustainable diet.

The situation may be somewhat different if we require hypothetical consent without idealization—that is, if we do not assume that the affected persons are fully informed and rational, but only consider how they would react if they were informed more narrowly about the intervention in question. Some would probably dissent to it due to a general dislike of government interference or forced behavior change. But it is hard to see why this should be invoked as a criterion of *moral* acceptability. (Note that we are not here, for once, concerned with moral psychology, but with a more genuinely moral question.) It would also make the consent requirement impossible to use in practice, since it would become pure guesswork to whom (and who many) a proposed intervention would be acceptable. Hence there is good reason to stick to the idealized version of the criterion, only requiring of a behavior intervention that those affected would endorse it if they were fully informed and fully rational.[3]

6.2 Goodbye to Liberalism?

The most common reason for finding behavior change interventions unacceptable is that they curtail people's autonomy. There seems to be a deep conflict between the fundamental liberal values of free choice, controlling one's own life, living in accordance with one's own values, etc., and the measures needed to combat climate change effectively. For the latter not only include large-scale social engineering and sweeping collective action. We have seen that it is probably also necessary to shape people's habits and choices beyond what can be achieved simply by informing them or presenting them with arguments. Moreover, the climate crisis turns out to be directly related to a class of behaviors that seem individually innocuous and have been seen as the epitome of private, individual life—household behavior, individual consumption and travel, etc. It thus puts pressure on the very idea of a private sphere that, according to liberal thinking, should not be subject to political regulation and perhaps not even to moral judgment (Jamieson & Paola, 2021).

Some might think that this is just the final nail in the coffin of liberalism. The climate crisis, they might say, only brings out its tensions and ramifications with definitive clarity. These have been evident for a long time. Concerns for economic and other substantial equality and the environment already put liberalism under pressure, before the greenhouse effect and its consequences became common knowledge. The climate crisis now just makes it crystal clear how private life and choices are—and always were—inextricably linked with the life and wellbeing of others and so not exempt from political and moral evaluation.

Yet there is another line of thought, similar in many of the details but less dismissive of liberalism as such, and so maybe less controversial. This is the idea that the core ideas of liberalism can be maintained even in the face of climate change, but that they support new practical conclusions in the new circumstances. Liberalism can itself *justify* curtailing the exercise of freedom. This can be said to follow from the celebrated *harm principle* put forward by Mill: people should be allowed to act as they wish unless their actions harm others (to a non-negligible degree; mere offense should be tolerated). Or as Mill himself put it in *On Liberty*: "[T]he only purpose for which power can be rightfully exercised over any member of a civilized community, against his will, is to prevent harm to others" (Mill, 1977, 223).

It should be noted that Mill himself did not take harm restriction to be the only possible justification of restrictions of liberty. He recognized further concerns for the general welfare as potentially relevant (Brink, 2022). Hence it seems that a paradigmatically liberalist view like Mill's is compatible with a degree of active welfare promotion. However, it is possible to argue that even a stricter, less social liberalism, which only accepts restrictions that can be shown to follow from the harm principle, can vindicate regulations and interventions of a seemingly radical sort. For it may be said that climate change has further revealed the degree to which human lives impact on one another. Actions previously thought to be innocuous have turned out to be not so. Spheres of life thought to be exclusively personal have turned out to be inextricably intertwined. Our knowledge of the sources and consequences of climate change should make us conclude that typical everyday behaviors, like car driving or meat consumption, pose a threat to other people's wellbeing, just as we have come to recognize that smoking in public places harms others significantly.

This is a cogent way of thinking, and it has the advantage of not pitching liberal and nonliberal views against each other. But while it maintains the justificatory basis of liberalism, it effectively relinquishes the ideal of liberty. This is quite a sacrifice. Many think that liberty has intrinsic value. And even if it has not, it is likely to have very broad and fundamental instrumental value, not least because it appears to be a precondition for exercising autonomy. If one does not believe that autonomy has intrinsic value either, then *its* broad instrumental value and very close correlation with wellbeing should be beyond dispute. And though the norms regulating the extent of the private sphere and the degree of separation between the private, collective, and public sphere are culturally relative, *some* kind of privacy seems to be very widely, if not universally, valued.

So although there is no need to deny that many allegedly private behaviors ought to be regulated, and that this is compatible with the basic principles of liberalism, we should also consider if more of its *substance*—the more specific liberal values, or liberal way of life—could be salvaged. This does seem to be the case. Climate change might also provide an occasion for clarifying and redetermining the content of liberalism. It can be argued that liberalism is not about unrestricted option freedom (like the freedom to choose between and consume many different products), anyway. Jamieson and Paola rightly note that

> ... actions that contribute to climate change and the Anthropocene are often not the kinds of actions whose privateness is particularly worth defending by liberal lights. They do not enhance diversity in desirable ways, and because they are typically relatively belief-independent acts of consumption they are not exactly paradigms of what we might hope for from the deliverances of practical reason. (Jamieson & Paola, 2021, 386)

It is debatable whether being a deliverance of practical reason by itself makes the action prudentially valuable. But it is a popular view among the liberal-minded, and so there is also, in any case, a point in reminding them that consumerist life is actually far from their ideal.

As for the positive substance of liberalism, the seminal texts of Humboldt and Mill which expound the liberal anthropology and value theory express a concern for diverse activities and experiences, continuous self-development, self-expression, room for experimenting with life forms and pleasures, creativity, and knowledge-seeking. Life with climate change restrictions need not reduce the potential for any of this. It may even enhance it, both because the green transition involves changes and challenges that call for precisely the qualities and activities that liberal axiology accords fundamental prudential value, and because the restrictions may help to free us from pressures and temptations that currently tend to curb our autonomy.

So it seems that the core content of liberalism can indeed be salvaged, even if we have to regulate many everyday behaviors. Perhaps some still worry that even leading a liberal lifestyle without excessive material consumption may not be sustainable in the long run. The liberal mindset, they might say, is still focused on growth, subtle diversification, and individual self-realization, and it is doubtful that there will be room for this in the future. The climate crisis will force us to move more closely together, share the resources, concentrate on meeting common challenges, and spend less on personal self-realization. This is not quite wrong. But it is not necessarily at odds with liberalism, which, in all its plausible versions, accept that liberal values must be balanced against other concerns, like those for future wellbeing and societal stability. Again, this would still entail some sacrifice. Yet more importantly, it can be argued that self-realization and diversification can be achieved very cheaply in terms of material *as well as* social resources. Personal growth can be decoupled if not completely, then to a very large degree, from other, more problematic kinds of growth.

Moreover, as emphasized by the *Bildung* tradition, autonomy is not the antithesis of communal life or collective and constrained action. They can go hand in hand. Collective projects can be simultaneously used as vehicles of self-realization and personal distinction. They do not require homogeneity or extreme conformity to succeed; quite to the contrary, successful collective action thrives on diversity (Gomez & Bernet, 2019). Likewise, there is little in the liberal notion of self-determination that implies that it is best exercised behind closed doors or that one is most authentically oneself in one's most private moments. Even if privacy is also a value that deserves to be protected, it should be distinguished from that of autonomy, which is not threatened by a stronger emphasis on joint activities.

To retain privacy while seeking to change even personal, everyday behaviors, Jamieson and Paolo suggest that we *moralize* rather than strictly regulate them, making people "their own censors" in such matters (2021, 387). They may have a point, especially if "moralizing" means not simply trying to *communicate* that the everyday behaviors in question are problematic but involves more serious attempts to make people *internalize* the negative valuations, for example, through role models and peer pressure. One piece of standard (but not necessarily classical) liberal thinking that may have to be abandoned is the idea that people can be expected to act rationally and apply their moral knowledge without external support.

Outright prohibitions of everyday behaviors related to personal hygiene, diet, mobility, and reproduction might indeed provoke backlash reactions, especially in countries like the United States, where many people are allergic to state interventions. Attempts to enforce such prohibitions would require intrusions into everyday life, perhaps also surveillance, which would itself constitute violations of privacy and be perceived as highly unpleasant. However, economic incentives like taxes and tariffs are also regulations, and it nevertheless seems very appropriate to apply them more strongly to behaviors like eating high on the food chain or driving, which Jamieson and Paolo mention among those who they think should be merely subject to moralizing (2021, 388).

Overall, I find it more promising to protect privacy by redefining its content and legitimate domains, just as is the case of autonomy. Yes, people should have a right to privacy. But there is no reason why privacy should be manifested by eating red meat or driving alone in a combustion engine car. Simply having time and space of one's own, being free from intrusions (which is not the same as being free from constraints, incentive schemes, or exposition to nudges) and surveillance may suffice.

6.3 Harm and Sustainability

The concern for avoiding harm can be used not only to justify interventions and restrictions on people's options, effectively curbing liberalism. It is more often used to justify restrictions on climate change action itself. That is the reason why assessing the potential impact on wellbeing on pro-environmental behavior change is an important element in climate ethics, as I argued in the introduction. The concern for not harming people seems, for example, to be built into what is perhaps the most influential definition of sustainability, that of the Brundtland Commission, which defined sustainable development as "meeting the needs of the present without compromising the ability of future generations to meet their own needs" (1987). It can be interpreted as a prohibition against imposing sacrifices on contemporary affluent people. As such, it could be justified by some kind of contractualism, especially contractualism of a broadly Kantian sort, which takes morality to be founded on principles all would, in principle, agree to. But ruling out sacrifices may, on the other hand, seem unrealistic (impossible, if the needs of future generations are also to be met), and unjust to further generations, who will in practice have to pay for keeping present people out of harm's way (Redclift, 2005; for a criticism of contractualism in the context of climate change, see Mulgan, 2015).

Of course, much depends on the understanding of the notoriously vague and ambiguous term "needs." Critics like Redclift seem to interpret it subjectively, assuming that people at different times or belonging to different groups may have different needs. This is in line with my own critical reception of the needs approach (see Section 3.5). However, it must be admitted that on an objectivist interpretation of needs, which is the standard view among proponents of the needs approach, the Brundtland criterion may allow that climate interventions do some harm to present people. Interventions may reduce their subjective wellbeing, as long as it remains above a certain threshold, and they are able to satisfy their basic needs. But this could still be said to be a constraint on climate action based on a concern for harm.

Hardcore consequentialists might argue that there is no limit to the amount of harm that may have to be endured by some people, if it is necessary to prevent future generations from suffering very large harms.[4] If the expected harms are very, very large and very certain, deontologist may eventually concur, because the negative consequences exceed the threshold beyond which consequentialist thinking should be allowed to hold sway (Moore, 2019). Philosophers of a deontological bent may further hold that present people deserve to experience reductions in their wellbeing both because of their partial responsibility for putting future generations in jeopardy and because it is most fair from the perspective of distributive justice. Yet it is probably not realistic to demand very large sacrifices in wellbeing from present people, both because of the risk of backlash effects and because wellbeing is a condition for effective action and altruism. Hence some kind of constraint on climate action along the lines of the Brundtland criterion may be pragmatically justified.

It is doubtful, however, whether this constraint will have to be applied very often. Just as the core values associated with the liberal way of life can likely be maintained even as we aim for ecological sustainability, wellbeing more generally can also be sustained despite necessary reductions in material consumption. The decoupling of wellbeing from specific consumption patterns and levels serves to deflate the issue of long-term sacrifice that has figured prominently in climate ethics.

It is unrealistic and unnecessary, however, to demand or expect that even a short-term significant reduction in wellbeing should be avoided. People do seem capable of undergoing, and willing to undergo, certain sacrifices provided they have enough reason to expect long-term improvement, experience the process as meaningful, and/or are inspired by strong role models. Because these factors also contribute to wellbeing, it is an open and quite subtle question to which extent there will actually be a reduction in wellbeing, and not just a change in its content, and so how fitting it is to speak of a sacrifice (although it *is* fitting to the extent that some aspects or elements of wellbeing are indeed sacrificed, the question is whether they are sufficiently compensated for). Anyhow, a strict prohibition against incurring harm is not justifiable even on pragmatic grounds. Proponents of the needs approach may be right in assuming that affluent people should be able to accept a reduction in their subjective wellbeing, though they are wrong in considering the reduction normatively irrelevant or insignificant.

6.4 Learning from the Pandemic

It is often said that social science must do without large-scale experiments. It is too costly, too risky, and also ethically inacceptable to rearrange the living conditions of large segments of the population in order to study the feasibility and consequences of such changes. Some radical political initiatives, like forced collectivization or the rapid introduction of market economies, are sometimes described as large social experiments. But it is then usually meant derogatively as a way of saying that the changes were implemented too rashly and with insufficient caution and realism.

However, the recent COVID-19 pandemic, and especially the measures taken to contain it, in many ways came close to a large-scale social experiment in swift collective behavior change. Of course, it was not deliberately designed, and the measures did not reflect any consistent methodological approach. There were no direct control groups, and the emergence of the virus and its different mutations were not caused by deliberate manipulation. Yet different policies were implemented in different countries and in different phases of the pandemic. Maybe it is more appropriate to compare it to large episodic natural phenomena, like an eclipse or supernova in astronomy, which can likewise compensate for the lack of real, controlled experiments. In any case, tragic though it was, the COVID-19 pandemic did provide extraordinarily comprehensive evidence that can significantly complement the results of small-scale psychological experiments and theoretical speculation.

We are here mainly interested in the question of acceptability of new behaviors imposed by authorities and sanctioned and enforced by informal peer pressure. But the change in policies during the pandemic further demonstrates the close link between acceptability and feasibility; measures were implemented and sustained to the extent that they were perceived as acceptable. Decisions to abandon them were of course often, at least partially, based on assessment of health risks and epidemic trajectories, but to no small degree also on the degree of public acceptance or resistance, as well as estimates of negative economic and social side-effects.

The overall lesson from the pandemic is mixed and seems to reiterate our main question about the possibility of smooth green transition, rather to answer it. We saw that quick and radical population-wide behavior changes—and reductions in resource-intensive activities—are indeed possible and need not take a large toll on people's wellbeing, at least in the short term. Living with restricted mobility and leisure opportunities can be surprisingly satisfying and elicit creative responses (Morse et al., 2021; Panarese & Azzarita, 2021). People are generally resilient and able to cope with change (Mau et al., 2022). But the experience from the later stages in the pandemic and its aftermath indicate that it can be difficult to sustain the changes in the long run (Hansen et al., 2021). Consumption and transportation behaviors quickly rebounded once restrictions had been lifted.

From the positive and negative experiences taken together, the following picture emerges. Continued acceptance and adherence depend on:

(1) Perceptions of the *usefulness* and *meaningfulness* of interventions. The apparent necessity and public benefit of the lockdown measures not only made it easier to comply but also made them feel less burdensome (Mau et al., 2022).
(2) Perceptions of one's efforts being *duly acknowledged*. This can be a matter of peer support, but also politicians expressing their support and appreciation (Mau et al., 2022).
(3) *Fairness* and creation of new *normality conditions*: new behaviors becoming a population-wide norm that applies equally to most social

groups. In the later stages of the pandemic, support for the lockdown measures waned partly because some groups stood out as being particularly strongly affected by them, and mutual solidarity crumbled (Johnson et al., 2023).

(4) *Variation*; options for development and activity; not feeling *stuck* in life (Gittings et al., 2021; Bernard & Slaby, 2023). The initial stages of the pandemic were for many a time of discovery, getting acquainted with and adapting to new circumstances. As the pandemic dragged on and the lockdown restrictions were retained, often inflexibly, the situation became increasingly perceived as claustrophobic, and the lack of a clear perspective for improvement of the situation felt depressing.

(5) Being *resourceful* and resilient from the outset. This may sound like a trivial observation. But people responded very differently to the pandemic, which tended to deepen inequalities, because those who were resourceful were only mildly, sometimes even positively, affected, whereas those who were financially, socially, or psychologically vulnerable were much more strongly affected and saw their situation aggravated. This reiterates the point that wellbeing matters for behavioral change also in the sense that people who already have a high level of wellbeing can more easily adapt to new conditions and cope with adversity. One important personal and collective resource is *trust*; the shift in collective mood, from wide adherence to the lockdown measures to fatigue and skepticism, was accelerated by a steady erosion of trust both in fellow citizens and authorities (Deslatte, 2020; Fang et al., 2023).

An important difference between the pandemic and the climate crisis is that the public and personal benefits of climate change measures are not so immediately apparent as those of virus mitigation. During the pandemic, many people felt immediately and personally at risk, or worried about the health of their close relatives. This difference is indeed significant but should not be exaggerated. The virus threat was also far from immediately visible or tangible; the perception of immediate danger was mainly induced by communication of abstract or distant states of affairs. It is likely that the perception of seriousness was prompted or enhanced by the fact that people were not only exposed to public warnings and recommendations, but that legal restrictions were also enacted. The pandemic was of course a still more acute crisis that did not trigger present bias or allowed for postponing action, as the climate crisis unfortunately does. Nevertheless, it is likely that (1)–(5), which both describe enablers of stable behavior change *and* conditions of conditions for acceptability, also apply, mutatis mutandis, to climate change actions. (5) is perhaps rather of condition for acceptance than for acceptability, as it pertains to the personal resources of those who must undergo the changes, including their threshold for toleration. But this is a good example of the more general point that the human interface matters: interventions may be ingenious and potentially powerful, but they are only as powerful as the people they are targeting allow them to be.

6.5 Realism, but Not Too Much: Balancing Feasibility Concerns and the Role of Hope

I have argued that a workable approach to green transition, and climate ethics generally, must be realistic. Moral obligations are constrained by practical limitations. The courses of actions that are worthy to consider in a situation as urgent as the climate crisis should not just be possible; they should be sufficiently likely to succeed. In other words, *feasibility* is an important concern. Yet although this seems indisputable, the implications are not always clear. What does it mean for a course of action to be "sufficiently likely to succeed"? What *is* feasibility? Realism tends to be invoked *ad hoc*, as realists often deem certain proposals unrealistic while assuming that it does make sense to aim for radical change in other domains (Pearson, 2023). The historical record shows that deeply entrenched institutions, the abolishment of which was considered utopian, like slavery, do eventually yield to moral pressure and may be swiftly overcome. The existence of tipping points beyond which new behaviors and ideas suddenly spread very swiftly (see Section 5.6) makes it difficult to estimate how realistic a certain course of action is. If we are not far from a tipping point, changing the *status quo* may not require large efforts, even though it seems so now.

The notion of feasibility is itself a source of epistemic uncertainty. It is applied to different things and in different ways. We speak of feasible actions, feasible solutions, feasible policies, and feasible goals. The threshold for feasibility and our means and capacities for assessing it differ widely, depending on which of these things we apply it to. A goal, like that of reducing emissions by 55 percent by 2030 or achieve climate neutrality by 2050 (as is the current EU target), may be feasible, because it could be achieved in very many ways, and some of these may be sufficiently likely to succeed. Yet specific actions or policies associated with the goal, for example, supporting and hoping for technological innovations, may not be feasible, because they are not sufficiently likely to achieve the goal.

There is no point in trying to regulate the use of the term "feasibility." However, for purposes of normative justification of inaction, a more restrictive use is preferable. We should avoid applying the term to single actions (or specific combinations of such actions), which are best left to specialized technical assessment of their workability. General goals and policies, on the other hand, may not be considered genuine plans or strategies at all (though quite a few so-called strategies, both in politics and business, seem to consist in little more than a statement of goals), and they may be most aptly characterized with the manifestly vague terms "realistic" and "unrealistic." A working definition that steers clear between these two extremes has been provided by Steele, who defines feasibility as *the probability that a project or multistage plan would be successfully realized if initiated in a particular way by some agent* (2021, 353). "Project" is a good term for the kind of multiply realizable, but still determinate and structured set of proposed tasks that may be deemed more or less feasible.

We should generally require that feasibility—the probability of successful realization—is sufficiently high if we are to spend resources on developing or carrying out a pro-environmental behavior change project. We should also require that goals or policies, while not themselves subject to feasibility assessment, only be adopted if we can devise multistage plans for their realization that are feasible in this sense. But there may be exceptions to this rule due to expected positive side-effects. A project may have a low chance of succeeding, but a large symbolic value, and the attempt to carry it out could be perceived as inspirational. We may also leave some leeway for subsequent experimentation and creativity, though it should still be constrained by considerations of overall likelihood of success. Merely assuming that creativity will fill in the blanks in a plan, because it has so often brought us unexpected solutions, is not acceptable.

As indicated above, there is a tendency to invoke feasibility considerations to halt or curb activities that may otherwise seem promising. It is noteworthy that feasibility concerns can themselves *reduce* feasibility. This is not just a theoretical possibility. It is a pervasive phenomenon. For example, it is presumably a central factor behind the tendency of politicians to hesitate with the implementation of climate change regulations (see Section 1.5, on the attitude-action gap and the negative feedback circuit between policymakers and citizens). The already complex decision-making is further complicated by the fact that policymakers should take into account expected public resistance but not make implementation challenges into a self-fulfilling prophecy. The latter is a serious risk not just because policymakers may overestimate the risk of resistance. They can prompt resistance by explicitly paying attention to the risk or showing concern for public worries, thereby creating an impression that the suggested policies will be difficult to accept.

That feasibility considerations can reduce feasibility is well known from the sports world, where athletes and teams are generally advised not to assess their chances of success objectively. If you are a contender facing an intimidating, hitherto undefeated opponent, your, in any case slight, chances of winning will be significantly better if you manage to believe that you are not so unlikely to win.

In the context of climate change action, the upshot is that being sensitive to acceptability issues can make it more difficult for actions to gain acceptance. This might to some extent justify deliberately playing down such issues. But it is still also the case that policymakers should consider the impact on wellbeing of their suggested policies *and* anticipate possible reactions. Realism should be maintained, but its complexity acknowledged, considering feedback and side effects. The aim should be to strike an admittedly delicate balance, making serious and comprehensive assessments and taking public worries seriously, but avoid framing changes in people's life circumstances as a threat to their wellbeing. Again, this otherwise difficult task is made easier by recognizing that the framing has a solid basis, since the changes *will* probably not impede people's wellbeing, apart from the inconveniences they may experience in the transition phase.

Analogous considerations pertain to individual plans for life changes. We should also be realistic in this regard, adhering to the Socratic maxim "know thyself" and only embarking on projects that match our actual skills and emotional nature. But when making such decisions, and sometimes even before, we should think also about the potentially demotivating effects of pondering feasibility too strongly, and the potentially motivating effects of believing that we are able to carry out the project. Calhoun remarks, "A … characterizing feature of difficult pursuits that seems connected with loss of motivation is the dependency of success on a large number of contingencies falling into place, including the success of multiple subplans" (2018, 75). Notice that this almost makes feasibility considerations inherently demotivating, because they—per definition—concern projects comprising of multiple subplans.

Again, this should not be taken to imply that we should not be concerned with the feasibility of our plans or aware of our personal limitations. But as in the policy case, our individual strategies should be calibrated to avoid the negative practical effects of ruminating over possible setbacks. We might even be justified in practicing a moderate degree of *self-deception*. As noticed above, making oneself believe that the chances of success are larger than the evidence indicates has long been a popular sports psychology technique. Flattering self-deception also fosters general resilience and power to act (though it can also reduce effort; for a nuanced survey of the positive and negative effects of cultivating positive illusions, see Schütz & Baumeister, 2017).

Some have described and justified *hope* in a way that seems to involve a kind of self-deception. Philip Pettit defines it as a preference for a prospect one believes to be possible, combined with the strategic formation of a tendency to act or think *as if* the desired prospect stood a good chance of obtaining (2004, 7). He justifies the adoption of such an attitude along the lines sketched above. Our emotional nature is such that keeping the negative evidence in mind will likely make us incapable of acting resolutely, even when this is our best option:

> [R]ationality does not always support … detached processing of fact in the cases where hope is relevant. What is rational for any one of us to do is what would best promote our goals and concerns, given the kinds of creature we are. But we are creatures of such a feeling, emotional kind that we cannot always serve our goals and concerns well under the image that our beliefs give us of the world we inhabit. Those beliefs may make disaster seem so likely, or they may wobble so uncertainly between the scenarios they hold out as probable, that they leave us in the slough of despondency, unable to act with decision or with any prospect of success.
>
> (2004, 9)

However, Petitt argues that adopting a more optimistic attitude is not really a case of self-deception:

> There is no self-deception of any kind involved in hope … For if the account given is correct, those who embrace hope need not indulge in any illusion or delusion about how things are. True, they set themselves to act and react as if things were otherwise than the evidence suggests they are or as if they were more firmly established than the evidence shows. But people can do this quite openly and honestly, being prepared to admit to themselves and others that for very good, pragmatic reasons—all the reasons that argue for cognitive resolve—they are refusing to expose themselves to the low or unstable tide of evidence;

> for current practical purposes, they are investing their confidence in a firmer, more encouraging prospect.
>
> (2004, 10f.)

Yet one can rightly ask if this is not indeed a case of, albeit innocuous, self-deception (Calhoun, 2018; Martin, 2013). It does involve an exercise of genuine self-knowledge because the adoption of the as-if attitude is based on an accurate assessment of one's emotional nature. But the attitude is itself a misrepresentation of one's situation. That is, rational to engage in such misrepresentation does not make it less deceptive. Hence hope is, thus understood, an example of how we may justifiably deceive ourselves to make our efforts more likely to succeed.

However, there may be ways of hoping, or otherwise adopting a more optimistic outlook than what seems to be supported by a sober assessment, that should not count as cases of self-deception. From the perspective of the *Bildung* tradition, it may be argued that a person is not simply what she presently is, but just as much what she is able to *become* if she aspires and develops in a certain way. Maybe I am not merely the person who easily succumbs to the temptation of shopping my blues away, but just as much the person who can successfully change this habit. In that case, an optimistic attitude may not be deceptive, but can be said to reflect a more complete knowledge of who I really am. It involves not a deliberate ignorance of parts of the available evidence, but rather adds further evidence which we might otherwise tend to ignore.

A similar line of thought has been pursued by Callard, who connects her idea of a person aspiring to change herself with the attitude of hope, without explicitly defining the latter. In line with the *Bildung* idea, she remarks, "We aspire by doing things, and the things we do change us so that we are able to do the same things, or things of that kind, better" (2018, 5). On Callard's view, aspiration is not generally immune to rational critique. It may be appropriate at some point for the aspiring person to conclude, or for others to try to persuade her, that she is not able to carry out the changes, that she doesn't "have it in her." Yet the early stages of the process should be exempted from such criticism (provided, I assume, that the project is not quite unrealistic). It should be allowed to properly take off and unfold; in the early stages it remains a genuinely open question whether the aspiring person is up to the task, and it can only be resolved in the very process of trying (Callard, 2018, 104). Something like this may also apply to the policy implementation cases. Prior to actual attempts at implementation, it may be difficult both to gauge the irritation it will cause and the resistance it will prompt *and* the capability improvements due to learning by doing.

Another, closely related way of entangling hope for self-deception is to deny that it needs to involve any belief in the probability of success. Calhoun suggests that we understand hope as involving a belief that success in a pursuit is possible, a preference that success is what actually materializes in the temporal unfolding of events, and a phenomenological idea of the determinate future whose content includes success. This is related to the *Bildung*-perspective because it urges us to not merely contemplate the future from *now*, but taking a view *from then*, projecting ourselves into a successful possible future, without thereby assigning it any specific probability. We simply "live under the idea" of a successful future (Calhoun, 2018, 86). This is not as such any justification of hope, but simply a description of a state that is only partially open to rational criticism, because it is not fully responsive to reasons. Our "phenomenological idea of the future" refers to our implicit *sense* of the future, which is an affective just as much as a cognitive state, rather than to what we believe about it or how we conceptualize it (Calhoun, 2018, 72).

However, Calhoun also indicates how the state of hope can be rationally justified. Apart from knowing that success is at least possible, the hoping person may find her pursuit *valuable enough* to assume risks of unredeemed costs (2018, 87). And this value assessment should consider both the importance of the goal and qualities intrinsic to the pursuit itself. The cost of failure may appear large, if it means that a lot of time has been wasted. From the perspective of our general, societal concern for curbing climate change, it might be rightly estimated as too large, if resources could have been spent on other projects that would have been more likely to succeed. Yet when viewed from a personal, but also from a group perspective, the costs may be mitigated if the efforts will likely be perceived as meaningful and rewarding, perhaps also as exemplary. It is easier to endeavor if what must be done is not itself unpleasant and the prospect of failure does not seem threatening.

How do these reflections on feasibility and hope relate to the issue of acceptability? They might seem to concern rather our experience and attitudes to change, which we discussed in Chapter 4. However, the wellbeing consequences and normative status of our attitudes to change are intimately connected with acceptability. The above reflections show that our own possibilities for modifying our attitudes and developing our capacities, and doing it as we go along, as part of the process of change, should be included in our assessment of whether and when a change in lifestyle may seem acceptable. Attitudes of commitment and hope make required changes easier to shoulder and so also accept. The reflections also show that inculcating or cultivating what could otherwise seem like excessively positive attitudes is *itself* acceptable, because it is both practically rational and compatible with appropriate self-knowledge and self-respect.

Besides, we have learned something about how to interpret the feasibility concern, which could otherwise lead us to abandon projects because of fears that they will not be acceptable to ourselves and others. The feasibility threshold should be relaxed because by endeavoring on a project, the people or organizations that carry it out may come to improve along the way. (This is also a kind of adaptation, more specifically of the coping sort.) Appeals to idealized, rational interests of people can be problematic because these are arguably not their *actual* interests, and it does not render them any more subjectively acceptable. Barring scruples about tinkering with people's personal development or self-formation, this criticism does not apply to cases where we have reason to expect them to adopt the rational interest in the very process of development. This is, once more, in accordance with the idea of *Bildung*: Inclination and duty, the subjective sense and objective standard of rightness, merge as one becomes a more complete version of oneself, by internalizing moral norms. The process is not one of simple imposition of norms from the outside, because the sense and actual livability of the norms are tested *in* the endeavor, which is intrinsically motivated.

Finally, regarding the worry that relaxing the feasibility threshold involves a kind of self-deception that may be justified because it serves the overall goal of stimulating climate action, but is not prudentially rational, it should be kept in mind that in the absence of significant alternatives, even projects with a quite low chance of success may be worth embarking on. This is especially true if these projects consist of activities that are not themselves particularly unpleasant.

6.6 Moral Aesthetics

Aesthetics plays an important role in making changes and interventions more acceptable. It does so on many levels and in many ways. An aesthetically pleasing environment is known to have considerable impact on people's wellbeing (Galindo & Rodriquez, 2000). There is also some, albeit less conclusive, evidence that perceptions of environments with low aesthetic value trigger emotions that support pro-environmental behavior (Li et al., 2022)—in other words, experiencing an environment as ugly or dirty can motivate pro-environmental action, just as fear sometimes can (McQueen, 2021). The latter must be considered a marginal phenomenon, however. The important connection between aesthetics and pro-environmental behavior change is positively supportive. Vegan and vegetarian dishes, sustainable travel destinations, sustainable fashion, housing, and public means of transport are much more likely to become popular if they are sufficiently aesthetic.

While awe and respect for nature hardly suffices to sustain pro-environmental behavior, the many positive experiences that contact with nature can provide may also help to make behavior change easier. They can provide both profound and more superficial, but still significant satisfaction, and compensate for the loss of other gratifying experiences. Both philosophical analyses of the aesthetics of nature and contemporary empirical research on the impact of nature exposures indicate that contact with nature support emotions and moods that are central to wellbeing, like calmness, attunement, or disinterested pleasure (Brady, 2015; Chang et al., 2024; Haybron, 2011).

This may be seen as a partial vindication of the view typical of the environmental humanities, viz. that we ought to cultivate a new relationship with nature. But it should be noted that the sceneries and environments that tend to be experienced as aesthetically pleasing and comforting need not be "natural" or "authentic" in any profound way; they will often be culture landscapes and can even be quite artificial. In the field of residential care, interventions understood as "biophilic"—aimed at meeting older people's assumedly innate tendency to seek connections with living things—are made *simulating* experiences of nature, which has proven very efficient at supporting older adults' wellbeing (Peters & Verderben, 2021). Nor should we assume that the kind of environment that results from effectively curbing greenhouse gas emissions will necessarily be aesthetically pleasing. Solar panels and wind turbines are widely perceived as impairing the visual quality of landscapes. Surely there is also a large potential for simultaneously providing positive aesthetic experiences and reducing emissions. But some of the most successful examples, like establishing public green spaces in cities, are blatantly artificial and anthropocentric.

Anyhow, the main point is that attempts to adopt a more pro-environmental behavior may be less strenuous if they provide aesthetic gratification. This is also central to the notion of *Bildung*, which aims not only at merging duty and inclination—making good feel good—but also to merge ethical judgment and aesthetic appreciation. The rationale behind this, which can otherwise seem like an unjustified fetichism of art, or exaggerated belief in the morally edifying power of the beautiful, seems to be that aesthetic appreciation or pleasure is the epitome of a rewarding experience more generally. Kierkegaard refers to everything that has a characteristic pleasurable experiential quality as "aesthetic" (Kierkegaard, 1992). This is not unreasonable. It is difficult to delineate the aesthetic more narrowly; and much of what is usually taken to contribute to hedonic, or more generally subjective, wellbeing does seem to have a broadly aesthetic quality. This is particularly vivid in those approaches to happiness and wellbeing that give emphasis to affective states.

More specific factors that contribute to making societal arrangements or interventions seem more or less acceptable are also at least partly aesthetic. We have already seen that people tend to prefer certain patterns or qualities of change and stability in life, much in accordance with aesthetic norms of continuity, well-roundedness, tension and resolution, unity in variety (Hekkert, 2006), etc. It is also noticeable that people prefer arrangements and environmental features that are convenient and comfortable, but that they can also accept and sometimes even want them to be challenging and complex. Things can get too easy, and people can maintain or discover the joy of mowing their lawn or brewing their coffee laboriously. This seeming contradiction accords with the aesthetic principle of self-determination (which is, for example, espoused by Schiller): for something to be aesthetically pleasing, it must be subject to certain rules and constraints and display tensions and recalcitrance, but these must not be imposed on it from without. In other words, it feels good to mow one's lawn or brew one's coffee as long as this is not done merely out of practical necessity but also for its own sake.

We can also speak of *moral aesthetics* in a more general sense. There arguably is a distinctive aesthetic pleasure associated with doing good or improving oneself, as pointed out by Kierkegaard (see Section 4.4). The wellbeing-conducive qualities of altruistic action that have been identified in many empirical studies, especially the "warm glow" experience that often accompanies it (Hartmann et al., 2017; Rhoads & Abigail, 2023; Taufik et al., 2015), can be characterized as aesthetic in a broad sense, as can moods and emotional states like attunement. Among the factors that contribute to make doing good feel good are appearing dynamic, powerful, moving or vivid, to oneself or others. These are also typical aesthetic qualities.

6.7 Unjustified Interventions?

A residual worry concerns the moral justification for the interventions. I have taken it for granted that it would be good and right if people changed their behavior, because this is needed to mitigate climate change. However, it has been argued persistently that we may not have any individual obligation to make low-emission lifestyle choices, or at least not to refrain from particular high-emission activities, like recreational SUV-driving (Sinnott-Armstrong, 2005; see also Nefsky, 2021). This skeptical view, or "inefficacy challenge" as it has come to be known, is troubling. Although the philosophical and scientific arguments that have been used to support it are unlikely to influence people's perceptions of the acceptability of climate action, arguments to the same conclusion, with roughly the same overall structure, are commonly used by ordinary people to justify inaction. People say things like "It does not make a difference if I give up driving a car, when the number of cars on the road keep increasing, anyhow." In debates about national climate policies, it is often claimed that the actions of a small country do not matter, as long as big countries like China and the United States do not significantly curb their emissions, or that high-emission production will just be taken over by other nations if local industries are forced to go green. It would be good if we could find compelling reasons to reject these skeptical thoughts.

Yet this seems much more difficult than should be expected. The standard reaction to people who think that their individual actions make no difference is to say that they do indeed make a difference, and that although it is admittedly very small, it is still significant, because it contributes to the overall problematic state of the world. Each person is responsible for her contribution to climate change and to climate change mitigation; and together, the many extremely small contributions add up to a hugely significant, better or worse, global outcome. Little brooks make great rivers, as the saying goes.

But this is the exactly the idea which the skeptical view challenges. Its proponents argue that our individual actions are so small and insignificant that they make *no* contribution to climate change. This is because global warming and climate change happen on a truly massive scale, and are processes already well under way:

> Global warming is [...] like a river that is going to flood downstream because of torrential rains. [If] I pour a quart of water into the river upstream [...] My

> act of pouring [a] quart [of water] into the river is not a cause of the flood. Analogously, my act of driving for fun is not a cause of global warming.
>
> (Sinnott-Armstrong, 2005, 298–9)

More specifically, if I refrain from some action, someone else somewhere on the globe will instantly do something similar in terms of emission production, and hence there will be practically no delay in reaching a problematic level of carbon dioxide concentration in the atmosphere (Nefsky, 2021, 204).

Besides, since the climate is a hugely complex, partly chaotic system, it is directly linked to neither any particular action nor any particular effect on the human world, the skeptics argue. Sinnott-Armstrong contends that *global warming*, to which an individual action can perhaps contribute a very tiny bit, must be distinguished from *climate change* as such:

> No storms or floods or droughts or heat waves can be traced to my individual act of driving. It is these climate changes that cause harms to people. Global warming by itself causes no harm without climate change. Hence, since my individual act of driving on that one Sunday does not cause any climate change, it causes no harm to anyone.
>
> (2005, 299)

Sinnott-Armstrong here seems to ignore the impact on wellbeing of more permanent factors like rising temperatures. But the general point remains: individual, emission-producing actions are not causally linked to any specific harmful effects and so cannot themselves be considered harmful or morally wrong.

The general thrust of the skeptical line of thought is that there is no ethical basis for demanding that people refrain from high-emission activities. But philosophical reasoning seems to support a further skeptical view, which also resonates with popular thinking: that ethically based requirements tend to be too *stern*. That is, if an ethical justification for criticizing high-emission activities can be found, it must then pertain indiscriminately to *all* such actions. But many think that "a little fun should still be allowed." And they may be right in this form if viewed from a feasibility perspective: Total abstention may be perceived as inacceptable, and insistence on this could compromise the attempt to bring about a generally more pro-environmental behavior. People may think that they do indeed have an obligation to moderate their consumption and act more pro-environmentally but fail to see that this should prohibit them from occasionally eating moderate amounts of meat or taking short recreational drives. And some may think that their generally pro-environmental lifestyle justifies such actions, which may be okay for them but not okay for people with a larger carbon footprint. Should person who hardly ever travels by plane not be allowed to make occasional recreational drives (Nefsky, 2021)?

However, if acts like recreational drives are morally wrong, then it seems that they are simply not permissible. People should refrain from them in any case, also if they are carried out on special occasions by individuals who lead an overall pro-environmental lifestyle. Hence it is not only that it can be difficult to explain why emission-producing actions are ethically problematic. We face a dilemma, because if we succeed in explaining this, then we thereby prevent ourselves from giving a sensible flexible response to different cases of emission production.

How should we respond to these kinds of skepticism? Fortunately, we do not need to resolve the issues very definitely. What matters to our purpose is merely that we can maintain, with sufficient firmness, that people generally ought to change their lifestyle and that they have moral reasons to generally abstain from emission-production actions. There is no need to demonstrate that standard, intuitively sound practices of praise, blame, encouragement, and discouragement are completely vindicated by normative theorizing. What we should aim for is a second-order philosophical justification of a first-order folk moral discourse that may not itself live up to strict philosophical standards.

I would not go so far as to insinuate that philosophers should hold back their skeptical considerations because these could be misunderstood or misused (since almost all philosophers involved in the discussion of the inefficacy challenge, including Sinnott-Armstrong, appear to agree that we should—"of course"—be concerned with and do a lot to curb global warming, and only ponder the role of individual acts and obligations, and *how* (but not really *whether*) to justify climate action). The arguments are unlikely to spill over to public discourse. But we should, in general, think carefully about how to apply philosophical normative reasoning to practical real-world issues, and be prepared to simplify and translate it before doing so. An example of such translation owns suggestion that we might (to some extent, and in some contexts) use values posited by objective wellbeing theory and notions of basic human needs for assessing the impact of climate change, even if this is not completely accurate and not warranted from the perspective of philosophical wellbeing theory (see Chapter 3).

In any case, a whole array of complementary responses to the inefficacy challenge are available. I sympathize with Broome's verdict that the more circumspect responses complicate climate ethics unnecessarily. Emission-producing actions are straightforwardly harmful in the morally relevant sense (if not by actually causing harm, then by carrying a *risk* of doing so, by being *expectedly* harmful) and by speeding up global warming and so increasing the risk of immediate climate changes (Broome, 2019). Again, we should keep in mind that contributing to global warming both increases the risk of harmful extraordinary events like droughts and floods, which Sinnott-Armstrong focuses on, and causes more permanent changes like a rise in mean temperature, which also have significant consequences for health and wellbeing as well as agricultural production (Cissé, 2022). *Pace* Sinnott-Armstrong, such changes should also count as very concrete manifestations of climate change.

Broome suggests, quite plausibly, that something like what has come to be known as the *social cost of carbon* (that is, the cost of the damages created by a carbon dioxide emission) is a reliable indicator of the expected harm caused by a single emission-producing act. Because this cost is independent of the behavior of other people, it is irrelevant what others may or may not do (Broome, 2019). The fact that another person somewhere else of the planet will instantly make a comparable contribution to global warming does not reduce the social cost of yours.

Hedden has argued in a similar vein that the existence of *buffers* in a system, which makes it less sensitive to individual acts, does not prevent such act from having—in some cases even large—long-term effects (Hedden, 2020).

However, if we should become convinced that the inefficacy challenge cannot be dismantled so straightforwardly, the more circumspect responses might be of help. They may create a false impression that climate ethics is more complicated than it actually is. But they are still good to have, since we had better build a strong *cumulative* case for refraining from emission-producing acts. If we can show that we have several independent reasons for criticizing such acts, this will make it much more difficult to evade the criticism.

First, the idea of individual acts somehow making a slight contribution, and adding up in a harmful way, is highly intuitive, and the skeptical case so dependent on controversial assumptions—or at least matters that are far from clear—that we should take its epistemic uncertainty into account as well. It is at best highly uncertain whether an individual act of, say, recreational driving does no harm whatsoever. The skeptical arguments are interesting, but while we discuss them back and forth, we had better look at our emission-producing activities with caution and show restraint.

Second, although both defenders and critics of the skeptical view tend to ignore possible side-effects of carrying out or refraining from the emission-producing acts—to avoid making an already quite complex assessment still more complicated—they all acknowledge that they may play a role, too. Hiller notes that there may be negative side-effects that enhance the wrongness of the action, for example, if driving will contribute to making this a habit (Hiller, 2011, 350). This point is reinforced by considering the mechanisms of habit formation (see Section 5.2). Nefsky thinks that the influence we may have by exposing others to pro-environmental behavior is "tiny" (2021, footnote 11). Sinnott-Armstrong's example of recreational driving is indeed constructed so that the act is not likely to influence others. But in real life, behaviors are noticed by others, and they are contagious, probably more so than Nefsky is willing to acknowledge. The mere knowledge that people go on recreational drives, and are okay with it, makes it seem normal and thus more acceptable.

Third, interventions targeting individual behavior do not have to be justified by individual obligations. They can be justified directly by our collective obligation to mitigate climate change. It is agreed on all hands that collective actions do make (or can be reasonably expected to make) a difference that is large enough to justify them. Even skeptics about individual obligation, like Sinnott-Armstrong, acknowledge a political responsibility to reduce emissions. Many of the interventions we are considering can be justified regardless of whether the individuals involved have an individual obligation to change their behavior. The interventions must not be understood as means for helping them to fulfill their obligations, but rather as the direct means for fulfilling our collective obligations. This may raise a worry about using people as means for achieving a larger good, that is, of instrumentalizing them. But that is a different question, related to the acceptability issues discussed above, and our discussion showed that the worry is hardly justified.

However, some of what I have, somewhat unorthodoxly, referred to as interventions may still seem to be in jeopardy. For they are not all large-scale collective actions. Must local actions that target specific individuals not be justified by specific obligations of those individuals? Could there be any justification for gently but firmly pushing your neighbor in a pro-environmental direction if she were not morally obliged to go there? Yes, there could. Provided that it is permissible to instrumentalize other people slightly for the sake of climate action,[5] and there is a chance that trying to influence your peers will make a difference, also because of the mechanisms of social contagion, you may do this without viewing it as helping your neighbor to fulfill any obligation of hers. In fact, it may be both practically *and* morally advantageous if attempts to make people to which we are closely related change their behavior are done *not* out of considerations of individual morality, since this may indeed be perceived as moralizing and inappropriate. Hence we need not worry that interventions aimed at bringing about behavior change cannot be justified, since they can be so independent of any issue of individual obligation.

It would still be preferable, however, if we could also show that individuals should view themselves as having some sort of individual obligation. Nefsky has suggested that this could be conceived in a less, but still sufficiently stringent, way, as a so-called "imperfect" obligation to avoid acting in ways that are *collectively* harmful *enough* of the time (Nefsky, 2021, 212). A different but similar approach is to focus not on the ethical acceptability of individual acts, but of overall *lifestyles* or activity patterns (like eating little meat or only rarely driving one's car) (Spiekermann, 2014). This also serves to avoid the demandingness challenge. On such a view, the person who generally acts pro-environmentally but occasionally indulges in some emission-producing activity can be justified in this because she still maintains a *lifestyle* of the right sort. This also tallies with the *Bildung* approach in focusing on general behavioral patterns and strategies rather than individual actions or dispositions to bring them about. A further step in this direction is to suggest that we can see an act as a problematic, and thus have a reason to refrain from it, even if it does not make any concrete, causal difference, if it can be said to contribute to a "vicious social pattern" (McPherson, 2021, 241).

But what about demandingness if individual acts *are* expectedly harmful, as I suppose they are? (I worry that people will be less motivated to refrain from emission-producing acts if these are likely to not do any difference, even though they can be said to contribute to a vicious social pattern. People will likely be tempted to use not doing any actual harm as an excuse for maintaining their old habits, regardless of how they think about other cases.) Fortunately, there are several lines of reply available. I noted above that philosophical, normative reasoning should not be directly applied to practical issues, but suitably translated. One way of doing this is to distinguish objective wrongness from blameworthiness (Driver, 2011, Ch. 5). Taking a recreational drive may be wrong, very strictly speaking, even for a person who pursues an otherwise environmentally friendly lifestyle. But we may not be entitled to blame her for it. And we probably should avoid doing so also to avoid the risk of backlash effects—perceptions of moral norms being too stern or applied too rigidly causing demotivation or resistance—and, more generally, to respect the (suitably modified) "ought implies can"-principle. One of the lessons learned from the COVID-19 pandemic is that compliance is difficult to achieve if obligations are perceived as being disproportionate or unfair.[6]

We can also avail ourselves of the fact that actual judgments of rightness are context-sensitive, depending on what one takes to be the relevant point of comparison (Driver, 2011, 51).

Although it may, strictly speaking, be wrong for the otherwise exemplary environment-friendly person to take a recreational drive, it may be such a small wrong, and so much better than what most other people tend to do, that we should not actually think of it as such. We can be stern in our overall philosophical assessment but lenient and pragmatic when it comes to our actual practices of praise and blame.

Notes

1 It might be said that paternalism could not be an issue here, since the actions in question are not motivated by the assumption that they will make the person interfered with be better off; they are done for the sake of the common good. It should be borne in mind, however, that it is quite possible that inducing green behavior will also benefit the person in question, both by protecting her from future harms from climate change and by enabling her to live a life that is less stressful and more satisfactory; and this may be part of the motivation for the actions, if not the main one.

2 I have encountered this view not least in research proposal reviews.

3 I am sympathetic to the objection that such highly idealized, hypothetical consent is not real *consent* at all—that it is too far removed from the actual attitudes of the person in question. I think it is really an embellished way to say that we should be allowed to assess the acceptability of an intervention regardless of what those affected may think of them. But I reckon that many would like to maintain some kind of consent requirement. Idealized consent is compatible with both my agent-neutral view and the Kantian rationale behind the proposal from Jun and colleagues.

4 I here ignore possible complications due to the "nonidentity problem" raised by David Parfit (1984), since utilitarianism does not accept the basic premise of the problem that an act can be wrong only if it harms some existing future person. For utilitarianism, it suffices to make an act wrong that it is likely to cause significant harm, regardless of whom will eventually suffer.

5 Here it should be remembered that all serious ethical theories, including that of Kant, allow that we use other people as instruments provided we not treat them solely as such.

6 This also illustrates why we should aim at adherence or concordance, rather than mere compliance with regulations and recommendations.

7

Changemaking: Gratifying Transitions and Strategies for Green Transition

In this last chapter, we shall finally address the "so what?"-question. What is to be done? How can we use knowledge about the conditions of wellbeing and behavior change to speed up green transition? I will try to translate the lessons that have emerged from my inquiries into recommendations aimed at both individuals and institutions. They are intended to be as concrete and straightforward as possible. But they concern the "how" of actions and initiatives rather than the "what." Finding and choosing the specific behaviors we would like to induce is a task for specialists in different fields. I assume that we already know many of them quite well. As laid out in Chapter 2, we will have to reduce our meat consumption, produce and shop more sustainably, recycle, reuse, and reassemble products like technology gadgets, live in smaller spaced homes, use more public transportation, and reduce our traveling distances. This is almost trivial. However, apart from the important task of devising specific and attractive ways to do so, it will also be an important part of our response to climate change to identify further relevant domains and affordances for pro-environmental behavior. The subsequent recommendations for how to go about pro-environmental behavior change should be applied not just to existing policies and goals, but also to those that may emerge in the coming years. The recommendations also serve as a kind of conclusion, summing up the main points as well as the arguments on which they rest.

(1) *Push on and persevere*, on all levels. Individuals, but also governments, local authorities, and institutions should unhesitantly continue their efforts to bring about pro-environmental behavior even in the face of setbacks and apparent failures. Of course, this does not mean that they should carry on with attempts that can be seen to be hopeless or far too costly compared to their likely outcome in terms of emission reductions.

Quite to the contrary. Another important maxim that follows directly from the realistic, unromantic approach advocated in this book is that we should be selective and seriously concerned with what works (see recommendation 2, below). The lesson is rather that perseverance is a condition for success in any case. Even interventions that can be assumed to work will only do so if they are maintained over a sufficient period. That solutions are realistic does not mean that they are easy or quick. But perseverance can overcome many obstacles and make even ambitious plans feasible. We should be realistic, but not self-effacing, and policymakers should be bold enough to maintain efforts in the face of temporary adversity. This attitude is justified by several points that have emerged throughout the book, especially the following:

i. *Repetition* is needed for behavior changes to stick. Habituation is not stopped by occasional, or even frequent, relapses into old habits. A very widespread mistake, in both in individual attempts to change and government interventions, is to give too much importance to setbacks instead of seeing them as inevitable bumps on the road. As argued in Chapter 6, feasibility concerns can be exaggerated and tend to create a negative feedback loop. In the individual case, the experience of not being able to maintain a new, desired behavior will often lead to self-doubt and lowered self-efficacy. This does not make the task easier, but still also does not make it impossible. We should keep doing the desired things even when we do not believe ourselves able to keep doing them. Here, the attitude-action gap can be an advantage: it is possible to achieve something one does not believe to be capable of. Yet it also shows how the gap can be bridged on the level of planning, strategy, and commitment: Knowing that repetition works can help us outsmart our inclinations and character flaws, even if we cannot conquer or correct them directly.

ii. It is possible that we may be close to a *tipping point* at which behavior changes more easily and rapidly (Section 5.5).

iii. Trying, and/or achieving intermediate goals, helps to succeed and establish new behaviors in the long run. Studies show that in cases where behavioral targets, like consistently recycling, appear hard to reach, a promising strategy is to support intermediate behavioral goals such as trying to recycle. Although initially falling short of the desired target, continuous attempts to recycle can eventually develop a culture of recycling in households (Issock et al., 2021).

iv. Willpower in a narrow sense may be a myth. We have no special power that enables us to directly resist temptations or put ourselves up by the bootstraps. But we can strengthen our ability to persevere by cultivating and regulating our emotions, redesigning our environment to make repeated efforts more natural and less burdensome, and making individual and especially collective commitments and providing mutual support (Section 5.6). Notably, if individuals *and* institutions *and* governments push on simultaneously, this mutually reinforces their chances of succeeding, even though there will be imperfection and setbacks on all three levels.

v. Even efforts that are unsuccessful with respect to their direct aim may contribute to change the overall environment and the perception of individuals, making them *generally* more inclined to accept and adhere to pro-environmental behavior changes. Even if I may not succeed in reducing my air traveling very much, my perceptible efforts to do so may influence those around me (Section 5.6).

vi. The problem of how to start and sustain a process of gradual improvement if the initial conditions are not favorable enough (Section 5.3) is best addressed by actually making and keeping things happening. By doing so, one both probes whether that conditions are in fact not favorable enough and contributes to improving them.

vii. Acceptability can, and often will, be achieved in the long run. There are many cases of ex post acceptance of interventions and changes that were at first quite unpopular. A striking example is the enactment of contestation pricing in the city of Stockholm in 2006. The traffic director later recalls that the closer they got to implementation, the more the drawbacks stood out, but now urges that policymakers fight through this "valley of political death," because support will quickly go up again when the positive effects become felt (Meyer, 2017). Interventions that do not lead to immediately noticeable improvements—even those that are initially experienced as unpleasant—are also likely to gain acceptance with time due to hedonic adaptation.

(2) *Work on the right levels and with the right types of behavior.* That is, work with either *quite specific* or *quite general* behaviors, and avoid the difficult middle range (Section 5.5). And aim at changing behaviors and dispositions that are relatively *superficial*, that is, not deeply rooted in people's identities, commitments, or cultural traditions.

Although there is reason to push on along a wide frontier, and on many levels, some targets are known to be more realistic and rewarding. On the one hand, in response to the huge impact of context, interventions should target specific behaviors in specific circumstances (e.g., making inhabitants of a specific city use its particular means of public transportation). On the other hand, because it is possible to foster general abilities to initiate, endure, and support change, and this is an important precondition for almost any more specific intervention, this should also be prioritized. It may be seen more as a general educational objective (supporting *Bildung* (Section 5.4)) but can be pursued specifically in the context of green transition, which provides a fertile ground for personal cultivation and development. Sunstein aptly remarks, again echoing the idea of *Bildung*: "Choice making is a muscle, and the ability to choose can be strengthened through exercise" (2019, 165). On a personal level, one can both set oneself very concrete tasks (e.g., only eating meat on certain occasions) and immerse oneself in a general process of development, embarking on a wholesale transformative experience (Section 4.1).

With regard to the actions of institutions and governments, information campaigns and traditional educational activities (that is, education aimed at fostering knowledge and attitudes) must be considered insufficiently effective means for short-term behavioral change. Apart from financial incentives, robust forms of behavioral design are more promising. Sustainable behavior should be made easy and natural, if not automatic, for example, by making renewable energy sources the default option for households and vegetarian or low-meat dishes the default—or only—option in corporate or public institution canteens. Almost anything that makes pro-environmental behavior slightly more convenient, changes the normality standards or support processes of habituation, is likely to work.

Targeting superficial rather than deeply rooted behaviors can be done by *changing the content rather than the form of an activity* (Section 2.4; 3.10). It is safer to change the ingredients in familiar dishes than to introduce new and exotic ones, to ensure that consumer goods like clothing is produced sustainably rather than curbing consumption or fashion trends as such, or to ensure that cars are driven by electric motors rather than forcing people to give up private transportation. The latter may not count as a case of behavior change at all but seem to be an example of the purely technology-driven, other-business-as-usual changes that proponents of technological fixes dream of (Section 1.4). Yet most technological solutions require some behavior change, though admittedly of the more superficial kind. For example, fans of combustion engines will be quick to emphasize that electric cars do not provide the same experience as their favorite gas guzzlers.

However, although it is a good guideline that one should aim at *incremental changes* in practices and try to not change their general form, the ability to adapt to more substantial changes should not be underestimated. A reason why policymakers should dare to be bold is that much of what might seem to be deeply rooted practices and preferences, even to people themselves, may likely turn out to be more superficial. A process of collective *Bildung* or aspiration also consists in testing, in practice, what our deeper values and commitments really are. We may not know that before embarking on the transitional process and learn how we react and what we are in fact capable of.

(3) Do not gamble on technological innovation but *assume and exploit wellbeing creativity* (Section 3.9). This means that we should worry less about costs and sacrifices in the longer term but pay close attention to the *immediate* and *local* reductions in wellbeing. People react adversely to felt inconveniences and abrupt lowering of material living standards, especially if they perceive themselves as being left worse off than others (Section 6.3). Yet over time, they not only adapt in the sense of becoming content with less but tend to find new ways of doing well—developing new rewarding activities or cultivating joys and pleasures that do not depend on the material resources that have now become scarcer. There are limits to the conditions under which this can fully compensate for material reductions, or for example, environmental degradation (Section 3.7). So this is not an argument for sitting back and letting climate change run its course, also because we should be concerned not only with maintaining sufficient wellbeing overall, but with everything that threatens to make people unnecessarily worse off (and also because climate change mitigation is itself unlikely to take a toll on wellbeing and may even contribute to enhancing it, due in part also to wellbeing creativity). It may, however, be a reason for not being extremely worried about the long-term impact of—suitably mitigated—climate change. That it will mean serious harm in the short- and mid-term is beyond discussion and so provides ample reason for taking action.

Wellbeing creativity is not just a factor to be passively reckoned with. It can and should be actively supported and employed. In one's personal life, it means that one should try to develop and cultivate one's potential for feeling good by doing good and try out activities and practices that may hold potential for wellbeing. On a collective level, we should support activities that make people acquainted with new and more sustainable sources of wellbeing. We have evidence from the COVID-19 pandemic (see Section 6.4) that people can learn to appreciate activities like local tourism, recreational walks, gardening, reading, or communal singing if they are urged to seriously try them out rather than merely considering them as abstract possibilities. That the learning process did not prevent people from lapsing back into old habits, for example, long-distance traveling, only shows that the initiatives should be supported by others, and not counteracted by pressures to resort to earlier consumption and transportation habits. Wellbeing creativity is a field where educational efforts can make a real difference. People who can get wellbeing out of different situations are more generally adaptable and can more easily take up new practices and technologies.

(4) *Imperfect solutions work and should be framed positively.* A main obstacle to green transition is that the available solutions are mostly quite imperfect and only partial. Despite widespread doubts even about that, electric cars have been shown to be significantly greener than those driven by combustion engines. However, while they reduce emissions with more than 60 percent in Europe (less in the United States, and still less in developing countries) (Negri & Biker, 2025), they are still far from carbon neutral when battery production is taken into account. This is of course no argument for sticking to combustion engines. But in practice it makes the switch to electric cars seem less compelling and existing transportation habits seem less problematic. This effect is often exacerbated by contemplating arguably superior, but not readily available, solutions like hydrogen cars.

That the available solutions are far from perfect should of course not be hidden from the public. But the solutions should be presented in ways that are sensitive to the fact that they are likely to be misperceived and misjudged. People should be supported in feeling that they do good by driving electric cars, employ small solar panels or engage in waste sorting even if the practices in question have clear limitations or negative side-effects. Learning to live, with imperfections, and getting them into the right perspective, is part of a *Bildung* process.

On the other hand, individuals and NGOs should also be wary that the legitimacy of imperfect solutions is not exploited for the wrong purposes, or that real alternatives are played down. We should opt for realism, but not too much (Section 6.5). Governments and companies may have interest in promoting an exaggerated realism and support solutions that are not the best available. Promoting biofuels or insisting on maintaining traditional modes of agricultural production with the justification that alternatives are worse and that it is part of a sensible "transition pathway" may be manipulation rather than just supportive framing of practices that are, all things considered, the best (but see Rial, 2024, for nuanced assessment of the use biofuels. Here, knowledge of what benefits and what does not benefit climate change really does matter. But this knowledge need not be widely dispersed among the public if it is possessed and acted upon by responsible key actors).

(5) *Be attentive to, and work with, subjective wellbeing.* Policies for green transition and particular types of sustainability interventions should, if possible, be tested for their impact on subjective wellbeing. Basic human needs and objective goods may be good proxies for wellbeing and sometimes easier to work with. Friendship and other close personal relationships, autonomy, achievement, etc., are often immediately connected with wellbeing: whatever supports the former tends to support the latter. However, attention to the subjective conditions for wellbeing is important, since allegedly objective goods can have markedly different effects in different contexts and on different people. Although it can generally be assumed that it is good to have access to nature, this does not mean that any initiative that improves access to nature will in fact be sufficiently beneficial to be worthwhile. The human interface is crucially important, and we might get more out of less by cultivating awareness and appreciation, strengthening people's ability to enjoy objectively less extensive or significant goods. There may be a lot to say for establishing further national parks, not least because of their role in maintaining biodiversity and a sustainable ecosystem. But potted plants, building-integrated vegetation, small gardens, and pocket parks may be more efficient when it comes to maintaining human wellbeing, regardless of what connoisseurs of scenic beauty may think about them. We might

quarrel philosophically about the value of authenticity, but when it comes to fostering subjective wellbeing, the potential of imitation and artificially produced goods is incontestable.

Focusing on subjective wellbeing also allows us to exploit the *substitutionability* of its forms, contents, and sources (see Section 3.6). Instead of lamenting the disappearance of things that have indeed been important to people's wellbeing, we should look for possible replacements. Often, important aspects of things and practices that people have been appreciating can be salvaged; sometimes they can even be strengthened. Bike riding can provide as much of thrill than driving a fast car. Friendship and other kinds of rewarding interpersonal relationships may profit from changes in life form due to concerns for curbing emissions (see Sections 2.5 and 2.6 on the expected changes in the built environment and forms of urban dwelling).

(6) *Seek and give social support.* Make green behavior change a *shared endeavor* (for example in your family or within a group of friends or colleagues) or try to emulate and follow someone with a more exemplary behavior. Commit yourself to collective activities, creating expectations and socials pressures that can keep on the track. This is a proven technique in fields like sports exercise and language learning. If you are already doing okay, then encourage other people around you to follow your example, not by lecturing or blaming them, but by displaying, tactfully but clearly, how you are able to feel good by doing good.

Because social relations are, if not intrinsically valuable, then at least very reliable and powerful sources of wellbeing, building such supportive social connections will likely be good, anyhow. Indeed, the positive side-effects may be so considerable that they should be considered mere side-effects at all. You might think that such social relations are not the best ones. Authentic togetherness is hardly achieved by recruiting others for a purpose; indeed, my recommendation could end up souring social relationships by instrumentalizing them. But this is a false opposition. Authentic togetherness is almost always bound up with projects and interests, be it sport, entertainment, culture, dating experiences, or the conduct of daily life. In friendships people usually play different roles and act as givers and takers of learning and inspiration.

(7) *Have no scruples about peer pressure and behavioral design.* Provided that the pressure one exercises on one's peers is mild, and mainly takes

the form of acting as an example and setting new standards, perhaps urging and encouraging new behaviors, it should give little reason for moral concern. The same goes for behavioral design in the public sphere. Behavioral designers should only require hypothetical (informed) consent by those affected (Section 6.1), though stakeholder analyses and use of participatory design may help to ensure that the values considered by designers are not too far from peoples' actual concerns, provided one has the resources needed for carrying them out. Perhaps the best practical justification is that behavioral design is already ubiquitous in so many fields. Think of speed bumps and road markings. Think of how streaming services give personalized recommendations based on users' viewing or listening histories, effectively outsourcing parts of cultural *Bildung*. Think of architecture promoting social interaction or inclusion (or, perhaps most vividly in earlier times, expressing reinforcing social hierarchies or power relations). There is nothing extraordinary, particularly paternalistic or dehumanizing, about using behavioral design to support pro-environmental behaviors.

(8) *But think about tone, style, and atmosphere—and work with aesthetics.* Many interventions are quite unoffensive and tend to be perceived as such. That is, they are hardly noticed as interventions or products of intentional design. But this also depends on how the interventions are presented. Aversions against paternalism, moralizing, and intrusions into the private sphere often reflect an *aesthetic annoyance* rather than a moral judgment. The morally right often appears aesthetically gratifying (see Sections 6.6), but it is also the other way around: the rightness of an activity can be obscured if it appears less aesthetically pleasing overall. We should nudge people not just in the sense of making it smooth and convenient, in practical or functional terms, for them to adopt new behaviors, but also in the sense of making it "nice" and aesthetically pleasing. An obvious example is that vegetarian dishes can be made less beige and more colorful; when presented beautifully, they may become almost irresistible. Another is that public transportation becomes more attractive if buses, trains, and the surrounding infrastructure are designed with a concern for aesthetics (see Cartenì & Cascetta, 2012, for concrete evidence for positive effects).

Some might doubt that the effect on climate change will be small enough to justify investments in the aesthetic aspects of transportation, infrastructure, or housing. But it should not be underestimated. Aesthetics is a powerful motivating factor. Features like rounded and interconnected shapes, the smell of grass or a bark-like surface have been shown to significantly facilitate the adoption of sustainable products (Nielsen et al., 2017). Still more importantly, because aesthetic improvements are almost certain to have a significant effect on wellbeing overall, they are very likely also justifiable overall, as are most attempts to work with wellbeing enhancement. There is little risk of efforts being wasted, because the means one uses is doing good, something which has value in any case.

But is the "seductive" power of aesthetics not a reason why we should have scruples about employing it (contrary to 6)? Is there not something particularly *disempowering* about influencing people's behavior by aesthetic means? Does the very smoothness, niceness, and apparent inoffensiveness of aesthetic interventions not make them particularly offensive? This is an understandable reaction, but it is based on a prejudice. There is no reason to assume that our responses to written messages, or even explicit arguments, are likely to be more "rational" and autonomous, or any less driven by emotions, concern for wellbeing, or aesthetic sensibility than our responses to an aesthetically designed product. Of course, we can think twice about whether it is right to follow a given advice, but we can also think twice about whether we should really prefer a service or product which is aesthetically attractive.[1] Remember that the interventions considered here are assumed to be ethically acceptable and politically legitimized. That is, they should be such that they would be accepted by the affected persons if these were fully rational and fully informed, and the overall decision to employ behavioral design should either have been made by democratic means or otherwise sanctioned by a politically legitimate authority.

As noticed in 6.1, we have no problem accepting something as intrusive and unappealing, if not ugly, as extensive taxpaying as legitimate and put up with it, simply because it has this kind of political legitimacy and is seen as a necessity. In comparison, employing aesthetics to facilitate the adoption of green solutions seems like a much smaller infliction; as indicated, it may be seen rather as an act of doing good. Contributing to an aesthetically pleasing environment should generally be seen as an important moral duty, because it is an important way of doing good, and because aesthetic annoyance and lack of aesthetic gratification constitute significant harms. Showing concern for aesthetics is also a way of showing respect for other people; this holds for both the public and the private sphere.

(9) *Work with variation!* Variation is a reliable means for keeping the good things alive and avoid that positive effects of changes wear off (Section 4.4). Some of this is closely related to aesthetics. "Unity in variety" is an aesthetic principle; lack of variation or diversity makes things less aesthetically pleasing and so less attractive. To use my standard example, vegetarian dishes should be just as varied as non-vegetarian, that is, there should be just as many and varied different dishes to choose between, and they should exhibit internal diversity as well. But there are other relevant aspects to variation. Interventions and pro-environmental practices should also be kept alive by being diversified and continuously developed. Mere repetition may be good at building lasting habits of a simpler sort. But when it comes to more general activities and ways of life, repetition breeds boredom, which in turn reduces motivation and engagement. We want people to take new ways of doing things to heart, spread them and develop them creatively; and we want them to notice progression and success, which reinforces their commitment. Hence if the resources available allow so, interventions should be continuously redesigned or at least presented in new ways. One of the many limitations of education for sustainability, as it is currently practiced in many places, is that it is made with little concern for progression and learners' cognitive and emotional development (see, e.g., Leifler & Dahlin, 2020); often the same (albeit relevant and instructive) cases and examples are presented repeatedly.

(10) *Embrace change!* We should, both individually and collectively, view the necessary transformations as an opportunity to initiate and exploit other kinds of change. There is no basis for fundamentally privileging living a changing over living a steady and uniform life (Sections 2.2 and 4.2). But there is ample evidence that living with and through changes *can* be more than sufficiently rewarding. People are often surprised to learn how they can thrive under changing, even apparently less friendly conditions, though there are of course also plenty of cases of negative surprises. Acknowledging that one has a considerable option space not so much for acts of consumption but for creatively redirecting or "refilling" one's life is

> a generally sound advice. Learning to live with and through change is an important part of building resilience and a rewarding investment also on a collective level.

It might be objected that although I pretend to be neutral as to whether we should prefer change or stability, I am in fact partial towards change and modernization (though this means something different in times of green transition than it did in times of industrial expansion). Am I not neglecting conservative views and the value of attachment to traditional forms of life? Am I also not forgetting that it is not just an abundance of consumption options, but also the exposure to, and pressure to try out or choose between, many different lifestyles and life directions, that can cause mental overload an anxiety?

No. Just as the core of liberalism may be salvaged, so may the core of conservatism. Conservatives are not so averse to change, nor should they be. In fact, they are the ones who champion the idea of changing to conserve. A recurrent theme in this book has been that seemingly large changes are not in fact so large or significant, because they only concern superficial aspects of life and do not alter its basic forms. This is supported by the framework for comparing different manifestations of change presented in Section 2.4. Embracing change can also mean daring to assume that it will not alter oneself and one's fundamental values, though it can also entail a preparedness to develop and perhaps even transform oneself.

Besides, a more sustainable life form is also a more stable one. The situation from which individuals and societies should contemplate to move away is not one that is characterized by respect for traditional values or a life of unperturbed leisure. It is correct that in contrast to many environmentalist thinkers, I do not envisage a single, final transition into a state of more sustainable, harmonious yet modest living, in which basic needs are provided for and further pursuit of happiness is stalled. This is not realistic—history will not end even if come to terms with climate change. Nor is it preferable, as humans do have a need for at least some kind of development, variation, and self-assertion. But I do envisage a situation in which personal development and diversification of life forms are decoupled from material growth and violently disruptive social transformations. Change will remain part of the human condition, but we can try and hope to domesticate it.

Coda: Living with Change in an Uncertain World

At the time of writing, there is much talk about how the world has become much more uncertain. The current age is commonly referred to as an "age of uncertainty." To what extent this is true is hard to say. Like the rate and depth of change in human affairs, existential uncertainty is also very difficult to measure. But in a way this only adds to the uncertainty, the subjective reality of which is hard to deny. The prospect of climate is only one, albeit a major, factor behind this new atmosphere of uncertainty and unpredictability. It has been known for quite some time, of course, as has its seriousness; but only recently has it begun to deeply affect people existentially and emotionally.

The new sense of fundamental uncertainty is also fueled by a recognition that history has not ended (*pace* Fukuyama, 1992). The belief typical of the 1990s and early 2000s, that Western-style liberal democracy was now unrivalled, that it would spread and remain a steady background for merely incremental changes and local variations, has turned out to be an illusion. Autocracies and populist democracies have re-emerged as serious contenders, and liberal democracies face both external and internal pressure. Recently, armed conflicts without precedent in the postwar era, and seemingly also without obvious solutions—notably in Ukraine and Gaza—threaten to destabilize the world order.

The current age is also marked by subjective uncertainties. The fragmentation of life and the disappearance of clear cultural norms have been characteristic of the whole late modern period; it has been felt by educated elites for more than 200 years and prompted several different responses. But it has arguably become more pervasive and problematic in recent years. The lack of a fixed life script and constant exposure to different, often contradictory, ideals and expectations have generated strong feelings of uncertainty especially among young (Lancet, 2022). Uncertainties about social connections, education, job security, and health coincide with increasing rates of anxiety and depression (Schweizer et al., 2023). The new uncertainty is not only an existential but also a health problem.

I have no suggestions for how to respond to this problem, which is of course also outside the scope of this book. It does, however, illustrate the importance of building resilience and strengthening people's, in this case especially young people's, ability to cope with change and embrace rather than fear the prospects of an open future, and underscores the importance of *Bildung* (Klausen & Mård, 2024). It is, moreover, a further example of the need for realistic approaches to societal challenges and a problem which we should not shy away from meeting with psychological techniques and behavioral design. Hence my point is mainly that the new uncertainty, if there is such a thing, highlights the centrality of how to live with, master, and work with change. To suggest that the current unstable world should be seen as a chance rather than as a threat may seem even more outrageous than suggesting it about climate change alone. But inasmuch that its unpredictability is a fact, we had better face up to it in a constructive fashion, focusing on the affordances for wellbeing and personal and societal development rather than the unavoidable risks, although, like in the case of climate change, we should take steps to mitigate and manage them as well.

Hence a case can be made for a moderate optimism, and for a hopeful attitude toward the future that not only entails reckoning with the good things that the future may have in store for us. That way of thinking, reasonable though it may be, is typical of the tendency to think of quality of life in terms of end states and outcomes rather the processes that lead to them, a tendency that is exemplified by the storage metaphor. It is a further noticeable sign of the neglect of change in normative theorizing that discussions of optimism versus pessimism in philosophy have revolved around the general, assumedly fixed, human condition: Is it a predicament or should it be considered good enough to make living worthwhile?[2] Strangely absent in such discussion are considerations of the possibility of working with and perhaps changing this very condition, or that it may have changed or be about to change due to external circumstances. We might hope that the future has good things in store for us. But we can be sure that it will be challenging. And so we can, with considerably more certainty, look to a future that will allow us to respond, adapt, aspire, and cultivate ourselves and so offer opportunities for gratifying transitions, regardless of where it will eventually lead us.

Notes

1 To what extent emotional responses can be said to be rational is, of course, a large and complex question. But the rationality of emotions has become increasingly appreciated in recent decades. And I suppose it is even less controversial to assume that our responses to texts and arguments are shaped by their rhetorical and aesthetic, and not just their epistemic, qualities.

2 For an influential, contemporary defense of pessimism, see Benatar (2017).

References

Acton, Harry B., and John W. N. Watkins. 2015. "Negative Utilitarianism." *Aristotelian Society, Supplementary Volumes* 37 (1): 83–114. https://doi.org/10.1093/aristoteliansupp/37.1.83.

Adams, Robert Merrihew. 1979. "Existence, Self-Interest, and the Problem of Evil." *Noûs* 13 (1): 53–65.

Adger, W. Neil, Jon Barnett, and Stacey Heath et al. 2022. "Climate Change Affects Multiple Dimensions of Well-Being through Impacts, Information and Policy Responses." *Nature Human Behavior* 6 (11): 1465–73. https://doi.org/10.1038/s41562-022-01467-8.

Ahn, Woo-kyoung, Nancy S. Kim, Mary E. Lassaline, and Martin J. Dennis. 2000. "Causal Status as a Determinant of Feature Centrality." *Cognitive Psychology* 41 (4): 361–416. https://doi.org/10.1006/cogp.2000.0741.

Albrecht, Gary L., and Patric J. Devlieger. 1999. "The Disability Paradox: High Quality of Life against All Odds." *Soc Sci Med* 48 (8): 977–88. https://doi.org/10.1016/s0277-9536(98)00411-0.

Alexander, Samuel. 2011. "Planned Economic Contraction: The Emerging Case for Degrowth." *Environmental Politics* 21 (3): 349–68. https://doi.org/10.2139/ssrn.1941089.

Alexandrova, Anna. 2017. *A Philosophy for the Science of Wellbeign*. Oxford: Oxford University Press.

Ali, Mir M., and Debra S. Dwyer. 2009. "Estimating Peer Effects in Adolescent Smoking Behavior: A Longitudinal Analysis." *The Journal of Adolescent Health: Official Publication of the Society for Adolescent Medicine* 45 (4): 402–8. https://doi.org/10.1016/j.jadohealth.2009.02.004.

Allik, Jüri, and Robert R. McCrae. 2004. "Toward a Geography of Personality Traits: Patterns of Profiles across 36 Cultures." *Journal of Cross-Cultural Psychology* 35 (1): 13–28. https://doi.org/10.1177/0022022103260382.

Alston, William P. 1993. *The Reliability of Sense Perception*. Ithaca, NY: Cornell University Press.

Antal, Miklós, and Jeroen C. J. M. Van Den Bergh. 2016. "Green Growth and Climate Change: Conceptual and Empirical Considerations." *Climate Policy* 16 (2): 165–77. https://doi.org/10.1080/14693062.2014.992003.

Appiah, Kwame A. 2008. *Experiments in Ethics*. Cambridge, MA: Harvard University Press.

Appiah, Kwame A. 2010. *The Honor Code: How Moral Revolutions Happen*. New York: W. W. Norton.

Arendt, Hannah. 2018. *The Human Condition*. Chicago, IL, & London: University of Chicago Press.

Aristotle. 2020. *The Nicomachean Ethics*. Translated by A. Beresford. London: Penguin Books.

Arler, Finn. 2014. "Miljøetik." [Environmental ethics]. In *Filosofisk Etik*, edited by Anne-Marie S. Christensen, 197–214. Aarhus: Aarhus University Press.

Armenta, Christina, Katherine Jacobs Bao, Sonja Lyubomirsky, and Kennon M. Sheldon. 2014. Is Lasting Change Possible? Lessons from the Hedonic Adaptation Prevention Model. In *Stability of Happiness: Theories and Evidence on Whether Happiness Can Change*, edited by K. M. Sheldon and R. E. Lucas, 57–74. Oxford: Elsevier Academic Press.

Arpaly, Nomy. 2010. "Comment." In *Meaning in Life and Why It Matters*, edited by Susan Wolf, 92–101. Princeton, NJ: Princeton University Press.

Bache, Ian, and Louise Reardon. 2016. *The Politics and Policy of Wellbeing: Understanding the Rise and Significance of a New Agenda*. Cheltenham: Edward Elgar.

Bader, Ralf M. 2016. "Conditions, Modifiers and Holism." In *Weighing Reasons*, edited by Errol Lord and Barry Maguire, 27–55. Oxford: Oxford University Press.

Bandura, Albert. 1977. "Self-Efficacy: Toward a Unifying Theory of Behavioral Change." *Psychological Review* 84 (2): 191–215. https://doi.org/10.1037/0033-295X.84.2.191.

Baumeister, Roy. 2019. "Self-Control, Ego Depletion, and Social Psychology's Replication Crisis." 10.31234/osf.io/uf3cn.

Baumeister, Roy F., and John Tierney. 2011. *Willpower: Rediscovering the Greatest Human Strength*. London: Penguin Press.

Bayne, Tim. 2010. *The Unity of Consciousness*. Oxford: Oxford University Press.

Beauvoir, Simone de. 2015. *The Second Sex*. New York: Vintage Classics.

Beethoven, L. van. 1997. *Heiligenstädter Testament*. Bonn: Beethoven-Haus (originally 1802).

Belshaw, Christopher. 2014. "What's Wrong with the Experience Machine?" *European Journal of Philosophy* 22 (4): 573–92. https://doi.org/10.1111/j.1468-0378.2011.00507.x.

Benatar, David. 2017. *The Human Predicament*. Oxford: Oxford University Press.

Bernhardt, Fabian, and Jan Slaby. 2023. "On Being Stuck: The Pandemic Crisis as Affective Stasis." *Phenomenology and the Cognitive Sciences* 22 (5): 1145–62. https://doi.org/10.1007/s11097-022-09855-1.

Besser, Lorraine L. 2020. *The Philosophy of Happiness: An Interdisciplinary Introduction*. London: Routledge. https://doi.org/10.4324/9781315283692.

Bissell, Paul, Carl R. May, and Peter R. Noyce. 2004. "From Compliance to Concordance: Barriers to Accomplishing a Re-framed Model of Health Care Interactions." *Social Science & Medicine* 58 (4): 851–62. https://doi.org/https://doi.org/10.1016/S0277-9536(03)00259-4.

Boström, Magnus, and Michele Micheletti. 2016. "Introducing the Sustainability Challenge of Textiles and Clothing." *Journal of Consumer Policy* 39 (4): 367–75. https://doi.org/10.1007/s10603-016-9336-6.

Bowen, Alex. 2019. "Green Growth or No Growth." *Grantham Research Institute on Climate Change and the Environment*, June 4, 2019. The London School of Economics and Political Science. https://www.lse.ac.uk/granthaminstitute/news/green-growth-or-no-growth/.

Brady, Emily. 2015. "Aesthetic Value, Nature, and Environment." In *The Oxford Handbook of Environmental Ethics,* edited by Stephen M. Gardiner, and Allen E. Thompson, 186–96. Oxford Handbooks: Oxford University Press. https://doi.org/10.1093/oxfordhb/9780199941339.013.17.

Bramble, Ben. 2016. "The Experience Machine." *Philosophy Compass* 11 (3): 136–45. https://doi.org/10.1111/phc3.12303.

Bramble, Ben. 2018. *The Passing of Temporal Well-Being*. London: Routledge.

Braudel, Fernand. 1982. *On History*. Chicago, IL: University of Chicago Press.

Brickman, Philip, and Donald T. Campbell. 1971. "Hedonic Relativism and Planning the Good Society." In *Adaptation-Level Theory*, edited by Mortimer H. Appley, 287–305. New York: Academic Press.

Brink, David. 2022. "Mill's Moral and Political Philosophy." In *The Stanford Encyclopedia of Philosophy* (Fall 2022 Edition), edited by Edward N. Zalta & Uri Nodelman. https://plato.stanford.edu/archives/fall2022/entries/mill-moral-political/.

Brittain, Amy. 2024. "Me Too Movement." *Encyclopedia Britannica*, April 19, 2024. https://www.britannica.com/topic/Me-Too-movement.

Britten, Nicky. 2014. "Concordance." In *The Wiley Blackwell Encyclopedia of Health, Illness, Behavior, and Society*, 299–302. Wiley: Oxford. https://doi.org/https://doi.org/10.1002/9781118410868.wbehibs038.

Broom, Douglas. 2022. "What's the Price of a Green Economy? An Extra $ 3.5 Trillion a Year." *World Economic Forum*, January 28, 2022. https://www.weforum.org/agenda/2022/01/net-zero-cost-3-5-trillion-a-year/#:~:text=The%20price%20of%20shifting%20to%20a%20green%20economy,down%20the%20cost%20of%20a%20green%20global%20economy.

Broome, John. 2019. "Against Denialism." *The Monist* 102 (1): 110–29. https://doi.org/10.1093/monist/ony024.

Brownstein, Michael, and Neil Levy. 2021. "Philosophy's Other Climate Problem." *Journal of Social Philosophy* 52 (4): 1–18. https://doi.org/10.1111/josp.12396.

Bruckner, Donald W. 2009. "In Defense of Adaptive Preferences." *Philosophical Studies* 142 (3): 307–24. https://doi.org/10.1007/s11098-007-9188-7.

Brülde, Bengt. 2015. "Well-Being, Happiness and Sustainability." In *Well-Being in Contemporary Society*, edited by Johnny H. Søraker, Jan-Willem Van der Rijt, Jelle de Boer, Pak-Hang Wong, and Philip Brey, 157–76. Cham: Springer International Publishing. https://doi.org/10.1007/978-3-319-06459-8_9.

Brundtland, Gro H. 1987. *Our Common Future: Report of the World Commission on Environment and Development*. Geneva, UN-Document A/42/427. http://www.un-documents.net/ocf-ov.htm.

Büchs, Milena, and Max Koch. 2019. "Challenges for the Degrowth Transition: The Debate about Wellbeing." *Futures* 105: 155–65. https://doi.org/https://doi.org/10.1016/j.futures.2018.09.002.

Buckwalter, Wesley. 2020. "Theoretical Motivation of 'Ought Implies Can.'" *Philosophia* 48 (1): 83–94. https://doi.org/10.1007/s11406-019-00083-7.

Buttrick, Nicholas R., Samantha J. Heintzelman, and Shigehiro Oishi. 2017. "Inequality and Well-Being." *Current Opinion in Psychology* 18: 15–20. https://doi.org/https://doi.org/10.1016/j.copsyc.2017.07.016.

Calhoun, Cheshire. 2018. *Doing Valuable Time: The Present, the Future, and Meaningful Living*. Oxford: Oxford University Press.

Callard, Agnes. 2018. *Aspiration: The Agency of Becoming*. Oxford: Oxford University Press. https://doi.org/10.1093/oso/9780190639488.001.0001.

Callard, Agnes. 2020. "Transformative Activities." In *Becoming Someone New: Essays on Transformative Experience Choice and Change*, edited by Enoch Lambert and John Schwenkler, 147–61. Oxford: Oxford University Press.

Carel, Havi, and Ian J. Kidd. 2020. "Expanding Transformative Experience." *European Journal of Philosophy* 28 (1): 199–213. https://doi.org/10.1111/ejop.12480.

Carens, Joseph H. 1996. "Realistic and Idealistic Approaches to the Ethics of Migration.": IMR: *The International Migration Review* 30 (1): 156. https://doi.org/10.2307/2547465.

Cartenì, Armando, and Ennio Cascetta. 2012. "The Value of Aesthetics in Public Transportation: A Quantitative Analysis of Perceived Quality for Railways Travel." Proceedings from 1st European Symposium on Quantitative Methods in Transportation Systems, Lausanne.

Carter, Evan C., and Michael E. McCullough. 2014. "Publication Bias and the Limited Strength Model of Self-Control: Has the Evidence for Ego Depletion Been Overestimated?" *Frontiers in Psychology* 5: 823. doi: 10.3389/fpsyg.2014.00823.

Caruso, Eugene M., Daniel T. Gilbert, and Timothy D. Wilson. 2008. "A Wrinkle in Time: Asymmetric Valuation of Past and Future Events." *Psychol Sci* 19 (8): 796–801. https://doi.org/10.1111/j.1467-9280.2008.02159.x.

Chang, Chia-chen, Brenda B. Lin, Xiaoqi Feng, Erik Andersson, John Gardner, and Thomas Astell-Burt. 2024. "A Lower Connection to Nature is Related to Lower Mental Health Benefits from Nature Contact." *Nature Scientific Reports* 14: 6705. https://doi.org/10.1038/s41598-024-56968-5.

Chen, Haiyan, Beisi Jia, and Sam S. S. Lau. 2008. "Sustainable Urban form for Chinese Compact Cities: Challenges of a Rapid Urbanized Economy." *Habitat International* 32 (1): 28–40. https://doi.org/https://doi.org/10.1016/j.habitatint.2007.06.005.

Christensen, Pil. 2020. "Selvfølgelig er Danmark et landbrugsland—og det skal det blive ved med at være." [Of course Denmark is an agrarian country—and so it shall remain]. *Information*, December 8, 2020. https://www.information.dk/debat/2020/12/selvfoelgelig-danmark-landbrugsland-ved-vaere.

Cissé, Guéladio, Robert McLeman, Helen Adams, Paulina Aldunce, Kathryn Bowen, Diarmid Campbell-Lendrum, Susan Clayton, Kristie L. Ebi, Jeremy Hess, Cunrui Huang, Qiyong Liu , Glenn McGregor, Jan Semenza, and Maria Cristina Tirado. 2022. "Health, Wellbeing, and the Changing Structure of Communities." In *Climate Change 2022: Impacts, Adaptation and Vulnerability*. (Contribution of Working Group II to the Sixth Assessment Report of the Intergovernmental Panel on Climate Change), edited by Hans-Otto Pörtner, Debra C. Robterts, Melinda M. B. Tignor, Elvira Poloczanska, Katja Mintenbeck, Andrés Alegría, Marlies Craig, Stefanie Langsdorf, Sina Löschke, Vincent Möller, Andrew Okem, and Bardhyl Rama, 1041–170. Cambridge: Cambridge University Press.

Cohen, Jeffrey J., and Stephanie Foote, eds. 2021. *The Cambridge Companion to Environmental Humanities*. Cambridge: Cambridge University Press. https://doi.org/DOI:10.1017/9781009039369.

Cohen, Maurie J. 2021. "New Conceptions of Sufficient Home Size in High-Income Countries: Are We Approaching a Sustainable Consumption Transition?" *Housing, Theory and Society* 38 (2): 173–203. https://doi.org/10.1080/14036096.2020.1722218.

Coveney, John. 2006. *Food, Morals and Meaning: The Pleasure and Anxiety of Eating*. 2nd Edition. London: Routledge.

Crimmins, Eileen M. 2021. "Recent Trends and Increasing Differences in Life Expectancy Present Opportunities for Multidisciplinary Research on Aging." *Nature Aging* 1 (1): 12–13. https://doi.org/10.1038/s43587-020-00016-0.

Crisp, Roger. 2006. *Reasons and the Good*. Oxford: Oxford University Press.

Crisp, Roger. 2021. "Well-Being." In *The Stanford Encyclopedia of Philosophy* (Winter 2021 Edition), edited by Edward N. Zalta. https://plato.stanford.edu/archives/win2021/entries/well-being.

Csikszentmihalyi, Mihaly. 1990. *Flow: The Psychology of Optimal Experience*. New York: Harper & Row.

Cummings, K. Michael, and Robert N. Proctor. 2014. "The Changing Public Image of Smoking in the United States: 1964–2014." *Cancer Epidemiology, Biomarkers & Prevention: A Publication of the American Association for Cancer Research, Cosponsored by the American Society of Preventive Oncology* 23 (1): 32–6. https://doi.org/10.1158/1055-9965.EPI-13-0798.

Cummins, Robert A. 1996. "The Domains of Life Satisfaction: An Attempt to Order Chaos." *Social Indicators Research* 38 (3): 303–32. https://doi.org/10.1007/BF00292050.

Darwall, Stephen. 2002. *Welfare and Rational Care*. Princeton, NJ: Princeton University Press. https://doi.org/doi:10.1515/9781400825325.

Decuypere, Anouk, and Wilmar Schaufeli. 2020. "Leadership and Work Engagement: Exploring Explanatory Mechanisms." *German Journal of Human Resource Management* 34 (1): 69–95. https://doi.org/10.1177/2397002219892197.

DeHaan, Cody R., and Richard M. Ryan. 2014. "Chapter 3—Symptoms of Wellness: Happiness and Eudaimonia from a Self-Determination Perspective." In *Stability of Happiness*, edited by Kennon M. Sheldon and Richard E. Lucas, 37–55. Cambridge, MA: Academic Press. https://doi.org/https://doi.org/10.1016/B978-0-12-411478-4.00003-5.

de Ridder, Denise T. D., Gerty Lensvelt-Mulders, Catrin Finkenauer, F. Marijn Stok, and Roy F. Baumeister. 2012. "Taking Stock of Self-Control: A Meta-Analysis of How Trait Self-Control Relates to a Wide Range of Behaviors." *Pers Soc Psychol* Rev. February, 16 (1): 76–99.

Deslatte, Aaron. 2020. "The Erosion of Trust during a Global Pandemic and How Public Administrators Should Counter It." *The American Review of Public Administration* 50 (6–7): 489–96. https://doi.org/10.1177/0275074020941676.

DeYoung, Colin G. 2015. "Openness/Intellect: A Dimension of Personality Reflecting Cognitive Exploration." In *APA Handbook of Personality and Social Psychology, Vol. 4. Personality Processes and Individual Differences*, edited by M. Mikulincer, P. R. Shaver, M. L. Cooper, and R. J. Larsen, 369–99. Washington, DC: American Psychological Association. https://doi.org/10.1037/14343-017.

Diekmann, Andres, and Peter Preisendörfer. 1998. "Environmental Behavior: Discrepancies between Aspirations and Reality." *Rationality and Society* 10 (1): 79–102. https://doi.org/10.1177/104346398010001004.

Doris, John M. 2002. *Lack of Character: Personality and Moral Behavior*. Cambridge: Cambridge University Press. https://doi.org/10.1017/CBO9781139878364.

Douthwaite, Richard. 1993. "The Growth Illusion." *Environmental Values* 2 (1): 79–80.

Doyal, Len, and Ian Gough. 1991. *A Theory of Human Need*. London: Red Globe Press.

Driver, Julia. 2011. *Consequentialism*. London: Routledge.

Dupuy, Jean-luc, Hélène Fargeon, Nicolas Martin-stpaul, François Pimont, Julien Ruffault, Mercedes Guijarro, Carmen Hernando, Javier Madrigal, and Paulo Fernandes. 2020. "Climate Change Impact on Future Wildfire Danger and Activity in Southern Europe: A Review." *Annals of Forest Science* 77 (2): 35. https://doi.org/10.1007/s13595-020-00933-5.

Ekdahl, David, and Lucy Osler. 2023. "Expressive Avatars: Vitality in Virtual Worlds." *Philosophy & Technology* 36 (2): 24. https://doi.org/10.1007/s13347-023-00628-5.

Elraz, Hadar, and Darren McCabe. 2023. "Invisible Minds: The Dominant Wellbeing Discourse, Mental Health, Bio-Power and Chameleon Resistance." *Organization* 30 (3): 490–509. https://doi.org/10.1177/13505084221145580.

Elster, Jon. 1983. *Sour Grapes: Studies in the Subversion of Rationality*. Cambridge: Cambridge University Press.

Encyclopedia Britannica. 2023. "Drive." November 16, 2018. https://www.britannica.com/topic/drive#ref49984.

Engelsen, Søren. 2022. "Wellbeing Competence." *Philosophies* 7 (2): 42. https://doi.org/10.3390/philosophies7020042.

Estrup, Eliza Jarl. 2021. Biodiversitet og klimaet. https://www.experimentarium.dk/klima/biodiversitet/.

European Environment Agency. 2020. "Climate Change Threatens Future of Farming in Europe." September 4, 2019. https://www.eea.europa.eu/highlights/climate-change-threatens-future-of.

European Environment Agency. 2023. "The Case for Public Participation in Sustainability Transitions." Publications Office of the European Union. https://www.eea.europa.eu/publications/the-case-for-public-participation.

Experimentarium. 2019. "Klimaforandringernes konsekvenser [Consequences of Climate Change]." August 18, 2019. https://www.experimentarium.dk/klima/klimaforandringernes-konsekvenser/.

Fagt, Sisse, Jeppe Matthiesen, and Anja P. Biltoft-Jensen. 2018. "Hvor meget kød spiser danskerne? Data fra statistikker og kostundersøgelser. [How Much Meat Do the Danes Consume?]." *DTU Fødevareinstituttet* 2018 (4): 1–6.

Fancourt, Daisy, Andrew Steptoe, and Alexandra Bradbury. 2022. "Tracking the Psychological and Social Consequences of the COVID-19 Pandemic across the UK Population: Findings, Impact, and Recommendations from the COVID-19 Social Study (March 2020—April 2022)." London: UCL. https://www.covidsocialstudy.org/final-report.

Fang, Guanfu, Tianyu Tang, Fang Zhao, and Ying Zhu. 2023. "The Social Scar of the Pandemic: Impacts of COVID-19 Exposure on Interpersonal Trust." *Journal of Asian Economics* 86: 101609. doi: 10.1016/j.asieco.2023.101609.

Farkas, Katalin. 2008. *The Subject's Point of View*. Oxford: Oxford University Press.

Fletcher, Guy. 2016. *The Philosophy of Well-Being: An Introduction*. 1st Edition. London: Routledge. https://doi.org/10.4324/9781315745329.

Fletcher, Guy. 2021. *Dear Prudence: The Nature and Normativity of Prudential Discourse*. Oxford: Oxford University Press. https://doi.org/10.1093/oso/9780198858263.001.0001.

Fletcher, Kate. 2019. *Wild Dress: Clothing and the Natural World*. Axminster: Uniformbooks.

Fletcher, Kate, and Mathilda Tham, eds. 2014. *Routledge Handbook of Sustainability and Fashion*. London: Routledge.

Frank, Robert H. 2020. *Under the Influence: Putting Peer Pressure to Work*. Princeton, NJ: Princeton University Press.

Fukuyama, Francis. 1992. *The End of History and the Last Man*. New York: The Free Press.

Galindo, Ma Paz, and José A. C. Rodriguez. 2000. "Environmental Aesthetics and Psychological Wellbeing: Relationships between Preference Judgements for Urban Landscapes and Other Relevant Affective Responses." *Psychology in Spain* 4 (1): 13–27.

García, Jorge H., Todd L. Cherry, Steffen Kallbekken, and Asbjørn Torvanger. 2016. "Willingness to Accept Local Wind Energy Development: Does the Compensation Mechanism Matter?" *Energy Policy* 99: 165–73. https://doi.org/https://doi.org/10.1016/j.enpol.2016.09.046.

Gardiner, Stephen M. 2006. "A Perfect Moral Storm: Climate Change, Intergenerational Ethics and the Problem of Moral Corruption." *Environmental Values* 15 (3): 397–413. http://dx.doi.org/10.3197/096327106778226293.

Gardiner, Stephen M. 2011. *A Perfect Moral Storm: The Ethical Tragedy of Climate Change*. Oxford: Oxford University Press.

Gauthier, Jeffrey A. 1997. "Schiller's Critique of Kant's Moral Psychology: Reconciling Practical Reason and an Ethics of Virtue." *Canadian Journal of Philosophy* 27 (4): 513–43.

Gilbert, Daniel. 2006. *Stumbling on Happiness*. New York: Alfred A. Knopf.

Gillebaart, Marleen. 2018. "The 'Operational' Definition of Self-Control." *Frontiers in Psychology* 9: 1231. doi: 10.3389/fpsyg.2018.01231.

Gillebaart, Marleen, and Denise T. De Ridder. 2015. "Effortless Self-Control: A Novel Perspective on Response Conflict Strategies in Trait Self-Control." *Social and Personality Psychology Compass* 9: 88–99. doi: 10.1111/spc3.12160.

Gittings, Lesley, Elona Toska, Sally Medley, Lucie Cluver, Carmen H. Logie, Nokubonga Ralayo, Jenny Chen, and Jane Mbithi-Dikgole. 2021. "'Now My Life Is Stuck!': Experiences of Adolescents and Young People during COVID-19 Lockdown in South Africa." *Global Public Health* 16 (6): 947–63. doi: 10.1080/17441692.2021.1899262.

Gladwell, Malcolm. 2002. *The Tipping Point: How Little Things Can Make a Big Difference*. New York: Back Bay Books.

Goethe, Johann Wolfgang v. 2011. *Wilhelm Meister's Apprenticeship and Travels*. Translated by Thomas Carlyle. Project Gutenberg. https://www.gutenberg.org/files/36483/36483-h/36483-h.htm#BOOK_II.

Gomez, L. E., and Patrick Bernet. 2019. "Diversity Improves Performance and Outcomes." *Journal of the National Medical Association* 111 (4): 383–92. https://doi.org/10.1016/j.jnma.2019.01.006.

Gonzalez-Arcos, Claudia, Alison M. Joubert, Daiane Scaraboto, Rodrigo Guesalaga, and Jörgen Sandberg. 2021. "'How Do I Carry All This Now?' Understanding Consumer Resistance to Sustainability Interventions." *Journal of Marketing* 85 (3): 44–61. https://doi.org/10.1177/0022242921992052.

Goss, Kristin A., and Philip Cook. 1996. "A Selective Review of the Social Contagion Literature." Terry Sanford Institute Working Paper, Duke University.

Gough, Ian. 2017. *Heat, Greed and Human Need*. Cheltenham: Edward Elgar Publishing.

Graeber, David, and David Wengrow. 2021. *The Dawn of Everything: A New History of Humanity*. London: Penguin.

Greif, Mark. 2017. *Against Everything: Essays*. New York: Vintage.

Guillén-Royo, Mònica. 2018. *Sustainability and Wellbeing: Human Scale Development in Practice*. London: Routledge. https://doi.org/10.4324/9781315762135.

Guyer, P. 1993. Kant and the Experience of Freedom. Cambridge: Cambridge University Press.

Haack, Susan. 2002. "Worthwhile Lives." *Free Inquiry* 22: 50–1.

Hagger, Martin S., Linda D. Cameron, Kyra Hamilton, Nelli Hankonen, and Taru Lintunen, eds. 2020. *The Handbook of Behavior Change*. Cambridge: Cambridge University Press.

Haidt, Jonathan. 2010. "Finding Meaning in Vital Engagement and Good Hives." In Comment in *Meaning in Life and Why It Matters*, edited by Susan Wolf, 92–101. Princeton, NJ: Princeton University Press.

Haines, Andy, and Kristie Ebi. 2019. "The Imperative for Climate Action to Protect Health." *N. Engl. J. Med* 380 (3): 263–73. doi: 10.1056/NEJMra1807873.

Hank, Petra, and Baltes-Götz Bernard. 2019. "The Stability of Self-Esteem Variability: A Real-Time Assessment." *Journal of Research in Personality* 79: 143–50. doi: 10.1016/J.JRP.2019.03.004.

Hansel, Tonya C., Leia Y. Saltzman, Pamela A. Melton, Tanisha L. Clark, and Patrick S. Bordnick. 2022. "COVID-19 Behavioral Health and Quality of Life." *Scientific Reports* 12 (1): 961. https://doi.org/10.1038/s41598-022-05042-z.

Hansen, Thomas, Thomas Sevenius Nilsen Marit Knapstad, Vegard Skirbekk, Jens Skogen, Øystein Vedaa, and Ragnhild Bang Nes. 2021. "Covid-Fatigued? A Longitudinal Study of Norwegian Older Adults' Psychosocial Wellbeing before and during Early and Later Stages of the COVID-19 Pandemic." *European Journal of Ageing* 19 (3): 463–73.

Harman, Elizabeth. 2015. "Transformative Experiences and Reliance on Moral Testimony." *Res Philosophica* 92 (2): 323–39. https://doi.org/10.11612/resphil.2015.92.2.8.

Hartmann, Patrick, Martin Eisend, Vanessa Apaolaza, and Clare D'Souza. 2017. "Warm Glow vs. Altruistic Values: How Important Is Intrinsic Emotional Reward in Proenvironmental Behavior?" *Journal of Environmental Psychology* 52: 43–55. 10.1016/j.jenvp.2017.05.006.

Hausman, D. M. 2015. *Valuing Health: Well-Being, Freedom, and Suffering*. Oxford: Oxford University Press.

Haybron, Daniel M. 2008. *The Pursuit of Unhappiness: The Elusive Psychology of Well-Being*. Oxford: Oxford University Press. https://doi.org/10.1093/oso/9780199545988.003.0015.

Haybron, Daniel M. 2011. "Central Park: Nature, Context, and Human Wellbeing." *International Journal of Wellbeing* 1 (2): 235–54.

Headey, Bruce W., and Alex J. Wearing. 1992. *A Theory of Subjective Well-Being*. Melbourne: Longman Cheshire.

Headey, Bruce, Ruud Muffels, and Gert G. Wagner. 2014. "Chapter 6—National Panel Studies Show Substantial Minorities Recording Long-Term Change in Life Satisfaction: Implications for Set Point Theory." In *Stability of Happiness*, edited by Kennon M. Sheldon and Richard E. Lucas, 99–126 Cambridge, MA: Academic Books. https://doi.org/https://doi.org/10.1016/B978-0-12-411478-4.00006-0.

Hedden, Brian. 2020. "Consequentialism and Collective Action." *Ethics* 130 (4): 530–54.

Heidegger, Martin. 2010. *Being and Time*. Translated by J. Stambaugh. Albany, NY: Suny Press.

Heinrich-Böll-Stiftung & ZOE - Institute for Future-Fit Economies, and Finanzwende Recherche, ed. 2022. "Making the Great Turnaround Work: Economic Policy for a Green and Just Transition." *Schriften zu Wirtschaft und Soziales*. Berlin: Heinrich-Böll-Stiftung e.V., vol. 27, 27.

Hekkert, Paul. 2006. "Design Aesthetics: Principles of Pleasure in Design." *Psychology Science* 48 (2): 157–72.

Henslin, James M., ed. 2010. *Sociology: A Down-to-Earth Approach*. 10th Edition. New York: Pearson Education, Inc.

Hiller, Avram. 2011. "Climate Change and Individual Responsibility." *The Monist* 94 (3): 349–68.

Horkheimer, Max, and Theodor W. Adorno. 2002. *Dialectic of Enlightenment*. Palo Alto: Stanford University Press. https://doi.org/doi:10.1515/9780804788090.

Horlacher, Rebecca. 2015. *The Educated Subject and the German Concept of Bildung*. London: Routledge.

Hosany, Sameer, and David Gilbert. 2010. "Measuring Tourists' Emotional Experiences toward Hedonic Holiday Destinations." *Journal of Travel Research* 49 (4): 513–26. https://doi.org/10.1177/0047287509349267.

Huizinga, Johan. 1949. *Homo Ludens: A Study of the Play-Element of Culture*. London: Routledge & Kegan Paul.

Hull, John M. 1990. *Touching the Rock: An Experience of Blindness*. London: Penguin.

Humboldt, Wilhelm v. 1967. *Ideen zu einem Versuch, die Grenzen der Wirksamkeit des Staats zu Bestimmen* [*Limits of State Action*]. Stuttgart: Reclam.

Humboldt, Wilhelm v. 2000. "Theory of *Bildung*" (1793). In *Teaching as a Reflective Practice: The German Didaktik Tradition*, edited by Ian Westbury, Stefan Hopmann, and Kurt Riquarts, 58–61. New York: Routledge.

Humboldt, Wilhelm v. 2015. *Gesammelte Schriften* XIV. 3. Abteilung: Tagebücher. Edited by Albert Leitzmann, and Bruno Gebhardt. Berlin: De Gruyter.

Hummel, Dennis, and Alexander Maedche. 2019. "How Effective Is Nudging? A Quantitative Review on the Effect Sizes and Limits of Empirical Nudging Studies." *Journal of Behavioral and Experimental Economics* 80: 47–58. https://doi.org/10.1016/j.socec.2019.03.005.

Hutton, Eric L. (trans.) 2014. *Xunzi: The Complete Text*. Princeton, NJ: Princeton University Press.

Ingendahl, Moritz, Dennis Hummel, Alexander Maedche, and Tobias Vogel. 2021. "Who Can Be Nudged? Examining Nudging Effectiveness in the Context of Need for Cognition and Need for Uniqueness." *Journal of Consumer Behavior* 20: 324–36. https://doi.org/10.1002/cb.1861.

Ip, Grace W.-M., Jing Chen, and Chi-Yue Chiu. 2006. "The Relationship of Promotion Focus, Need for Cognitive Closure, and Categorical Accessibility in American and

Hong Kong Chinese University Students." *Journal of Creative Behavior* 40: 201–5. https://doi.org/10.1002/j.2162-6057.2006.tb01273.x.

Isham, Amy, Birgitta Gatersleben, and Tim Jackson, 2021. "Materialism and the Experience of Flow." *J Happiness Stud* 22 (4): 1745–68. https://doi.org/10.1007/s10902-020-00294-w.

Issock, Paul Blaise Issock, Mercy Mpinganjira, and Mornay Roberts-Lombard. 2021. "Trying to Recycle Domestic Waste and Feelings of Guilt: A Moderated Mediation Model Applied to South African Households." *Sustainable Production and Consumption* 27: 1286–96.

Jacobs, Michael. 2013. "Green Growth." In *The Handbook of Global Climate and Environment Policy*, edited by Robert Falkner, 197–214. Oxford: Wiley. https://doi.org/https://doi.org/10.1002/9781118326213.ch12.

Jamieson, Dale. 2014. *Reason in a Dark Time: Why the Struggle against Climate Change Failed—And What It Means for Our Future*. New York: Oxford University Press.

Jamieson, Dale, and Marcello D. Paola. 2021. "Climate Change, Liberalism, and the Public/Private Distinction." In *Philosophy and Climate Change*, edited by Mark Budolfson, Tristram McPherson, and David Plunkett, 370–96. Oxford: Oxford University Press. https://doi.org/10.1093/oso/9780198796282.003.0017.

Jensen, Peter B., Linda Nhu Laursen, and Louise Møller Haase. 2021. "Barriers to Product Longevity: A Review of Business, Product Development and User Perspectives." *Journal of Cleaner Production* 313: 127951. https://doi.org/https://doi.org/10.1016/j.jclepro.2021.127951.

Joewono, Tri B., Mohamed Y. F. Wirayat, Prawira F. Belgiawan, I. G. A. Andani, and Clint Gunawijaya. 2023. "Users' Preferences in Selecting Transportation Modes for Leisure Trips in the Digital Era: Evidence from Bandung, Indonesia." *Sustainability* 15 (3): 2503. https://doi.org/10.3390/su15032503.

Johnson, Stephanie, Steven Roberts, Barbara Prainsack, Sarah Hayes, Amelia Fiske, Federica Lucivero, Stuart Mclennan, Amicia Phillips, and Gabrielle Samuel. 2023. "Understanding Pandemic Solidarity: Mutual Support during the First COVID-19 Lockdown in the United Kingdom." *Public Health Ethics* November, 16 (3): 245–60. https://doi.org/10.1093/phe/phad024.

Jones, Christopher R., and J. Richard Eiser. 2010. "Understanding 'Local' Opposition to Wind Development in the UK: How Big Is a Backyard?" *Energy Policy* 38 (6): 3106–17. https://doi.org/https://doi.org/10.1016/j.enpol.2010.01.051.

Jones, Nicola. 2018. "How to Stop Data Centres from Gobbling Up the World's Electricity." *Nature* 561 (7722): 163–6. https://doi.org/10.1038/d41586-018-06610-y.

Jordan, Karen, and Kristján Kristjánsson. 2017. "Sustainability, Virtue Ethics, and the Virtue of Harmony with Nature." *Environmental Education Research* 23 (9): 1205–29.

Jun, Gyuchan T., Fernando Carvalho, and Neil Sinclair. 2018. "Ethical Issues in Designing Interventions for Behavioral Change." In *Proceedings of the Design Research Society 2018* (Limerick, Ireland), edited by Cristiano Storni et al., 112–24. https://doi.org/10.21606/drs.2018.498.

Kagan, Shelly. 2012. *Death*. New Haven, CT: Yale University Press.

Kagan, Shelly. 2019. *How to Count Animals, More or Less*. Oxford: Oxford University Press. https://doi.org/10.1093/oso/9780198829676.001.0001.

Kahneman, Daniel, and Angus Deaton. 2010. "High Income Improves Evaluation of Life but Not Emotional Well-Being." *Proc Natl Acad Sci U S A* 107 (38): 16489–93. https://doi.org/10.1073/pnas.1011492107.

Kahneman, Daniel, Alan B. Krueger, David A. Schkade, Norbert Schwarz, and Arthur A. Stone. 2004. "A Survey Method for Characterizing Daily Life Experience: The Day Reconstruction Method." *Science* 306: 5702, 1776–80.

Kamtekar, R. 2004. "Situationism and Virtue Ethics on the Content of Our Character." *Ethics* 114 (3): 458–91.

Kasser, Tim. 2002. *The High Price of Materialism*. Cambridge, MA: The MIT Press. https://doi.org/10.7551/mitpress/3501.001.0001.

Kasser, Tim. 2017. "Living Both Well and Sustainably: A Review of the Literature, with Some Reflections on Future Research, Interventions and Policy." *Philosophical Transactions of the Royal Society A: Mathematical, Physical and Engineering Sciences* 375 (2095): 20160369. https://doi.org/doi:10.1098/rsta.2016.0369.

Kelly, Daniel. 2022. "Two Ways to Adopt a Norm: The (Moral?) Psychology of Internalization and Avowal." In *The Oxford Handbook of Moral Psychology* (online edition), edited by Manuel Vargas, and John M. Doris, 285–309. Oxford: Oxford University Press.

Kierkegaard, Søren A. 1992. *Either/Or. A Fragment of Life*. Translated by Alastair Hannay. London: Penguin (E/O).

Kierkegaard, Søren A. 2013. *Søren Kierkegaards Skrifter*. Edited by N. J. Cappelørn et al. Copenhagen: Søren Kierkegaard Forskningscenteret. http://sks.dk/forside/indhold.asp (SKS).

Kim, Heejung S., and Aimee Drolet. 2003. "Choice and Self-Expression: A Cultural Analysis of Variety-Seeking." *Journal of Personality and Social Psychology* 85 (2): 373–82. https://doi.org/10.1037/0022-3514.85.2.373.

Kim, Heejung S., and Hazel R. Markus. 1999. "Deviance or Uniqueness, Harmony or Conformity? A Cultural Analysis." *Journal of Personality and Social Psychology* 77 (4): 785–800. https://doi.org/10.1037/0022-3514.77.4.785.

Kind, Amy. 2020. "What Imagination Teaches." In *Becoming Someone New: Essays on Transformative Experience, Choice, and Change* (online edition), edited by Lambert Enoch, and John Schwenkle, 133–46. Oxford: Oxford University Press. https://doi.org/10.1093/oso/9780198823735.001.0001.

Kirkland, Giles. 2020. "Car Culture and Climate Change: Personal and Systemic Solutions." *Millennium Alliance for Humanity and the Biosphere*. https://mahb.stanford.edu/blog/car-culture-and-climate-change-systemic-and-personal-solutions/.

Klafki, Wolfgang. 2000. "The Significance of Classical Theories of Bildung for a Contemporary Concept of Allgemeinbildung." In *Teaching as a Reflective*

Practice: The German Didaktik Tradition, edited by Ian Westbury, Stefan Hopmann, and Kurt Riquarts, 85–108. Mahwah, NJ: Lawrence Erlbaum.

Klausen, Søren H. 2008. "The Phenomenology of Propositional Attitudes." *Phenomenology and the Cognitive Sciences* 7: 445–62. https://doi.org/10.1007/s11097-007-9081-z.

Klausen, Søren H. 2013. "Losing Belief, While Keeping Up the Attitudes: The Case for Cognitive Phenomenology." In *New Essays on Belief*, edited by Nikolaj Nottelmann, 188–208. London: Palgrave Macmillan. https://doi.org/10.1057/9781137026521_10.

Klausen, Søren H. 2016. "Happiness, Dispositions and the Self." *Journal of Happiness Studies* 17 (3): 995–1013. https://doi.org/10.1007/s10902-015-9628-6.

Klausen, Søren H. 2018a. *På sporet af din lykke* [In Pursuit of Happiness]. Copenhagen: Akademisk Forlag.

Klausen, Søren H. 2018b. "Ethics, Knowledge, and a Procedural Approach to Wellbeing." *Inquiry* 66 (1): 1–17. https://doi.org/10.1080/0020174X.2018.1529619.

Klausen, Søren H. 2018c. *Kierkegaard. Educating for Authenticity*. Cham: Springer.

Klausen, Søren H. 2020a. "Nihilism, Depression and Wholeheartedness: Metacognitive Strategies in 19th Century Literature." In *Faces of Depression in Literature*, edited by Josefa R. Velasco, 71–85. New York: Peter Lang.

Klausen, Søren H. 2020b. "The Essential Tension in Phenomenal Consciousness." *Philosophical Papers* 49 (1): 159–90.

Klausen, Søren H. 2022. "Lettere sagt end gjort. Udfordringer i samspillet mellem filosofisk teori og empiri" [Easier Said than Done. Challenges to the Interplay between Philosophical Theory and Empirical Evidence]. In *Filosofi og empiri*, edited by Anita N. Holm, 11–31. Aalborg: Aalborg University Press.

Klausen, Søren H., and Nina Mård. 2024. "For Life, Knowledge and Work: Rationale and Aims of Crosscurricular Teaching and Learning." In *Across and Beyond School Subjects. A Framework for Crosscurricular Didactics*, edited by S. H. Klausen and Nina Mård, 19–32. London: Routledge.

Klausen, Søren H., Søren Engelsen, Regina Christiansen, and Jakob Emiliussen. 2020. "Elderly Well-Being and Alcohol: A Tricky Cocktail." *International Journal of Qualitative Methods* 19: 1609406920931687. https://doi.org/10.1177/1609406920931687.

Klausen, Søren H., Regina Christiansen, Jakob Emiliussen, Lejla Hasandedic-Dapo, and Søren Engelsen. 2021. "The Many Faces of Hedonic Adaptation." *Philosophical Psychology* 35 (2): 253–78. http://dx.doi.org/10.1080/09515089.2021.1967308.

Kollmuss, Anja, and Julian Agyeman. 2002. "Mind the Gap: Why Do People Act Environmentally and What Are the Barriers to Pro-Environmental Behavior?" *Environmental Education Research* 8 (3): 239–60. https://doi.org/10.1080/13504620220145401.

Kristal, Ariella S., and Ashley V. Whillans. 2020. "What We Can Learn from Five Naturalistic Field Experiments That Failed to Shift Commuter Behaviour?." *Nature Human* Behavior 4 (2): 169–76. doi: 10.1038/s41562-019-0795-z.

Krys, Kuba, John M. Zelenski, Colin A. Capaldi, Joonha Park, Wijnand van Tilburg, Yvette van Osch, Brian W. Haas, Michael H. Bond, Alejandra Dominguez-Espinoza, Cai Xing, David O. Igbokwe, Anna Kwiatkowska, Magdalena Luzniak-Piecha, Martin Nader, Muhammad Rizwan, Zichen Zhu, and Yukiko Uchida. 2019. "Putting the 'We' into Well-Being: Using Collectivism-Themed Measures of Well-Being Attenuates Well-Being's Association with Individualism." *Asian Journal of Social Psychology* 22 (3): 256–67. https://doi.org/https://doi.org/10.1111/ajsp.12364.

Kuck, Gretchen, and Gary Schnitkey. 2021. "An Overview of Meat Consumption in the United States." *farmdoc daily* 11: 76. Department of Agricultural and Consumer Economics, University of Illinois at Urbana-Champaign.

Kurcharski, Adam. 2020. *The Rules of Contagion: Why Things Spread—and Why They Stop*. London: Welcome Collection.

Lambert, Enoch, and John Schwenkler, eds. 2020. *Becoming Someone New: Essays on Transformative Experience, Choice, and Change*. Online Edition. Oxford: Oxford University Press. https://doi.org/10.1093/oso/9780198823735.001.0001.

Lancet, the. 2022. "An Age of Uncertainty: Mental Health in Young People." *The Lancet* 400 (10352): 539.

Łaszewska, Agata, Timea Helter, and Judit Simon. 2021. "Perceptions of Covid-19 Lockdowns and Related Public Health Measures in Austria: A Longitudinal Online Survey." *BMC Public Health* 21 (1): 1502.

Leifler, Ola, and Jon-Erik Dahlin. 2020. "Curriculum Integration of Sustainability in Engineering Education." doi: 10.1108/IJSHE-09-2019-0286.

Leung, Kwok, and Michael M. Morris. 2010. "Culture and Creativity: A Social Psychological Analysis." In *Social Psychology and Organizations*, edited by David D. Cremer, J. Keith Murnighan, and Rolf V. Dick, 371–95. Boca Raton, FL: Taylor & Francis.

Li, Hong. 1977. *China's Political Situation and the Power Struggle in Peking*. Hong Kong: Lung Men Press.

Li, Hui, Chang You, Jin Li Mei Li, Min Tan, Guanfei Zhang, and Yiping Zhong. 2022. "'Influence of Environmental Aesthetic Value and Anticipated Emotion on Pro-Environmental Behavior': An ERP Study." *International Journal of Environmental Research and Public Health* 19 (9): 5714. https://doi.org/10.3390/ijerph19095714.

Liao, Bin, Lin Li, and Zhao Yang. 2022. "Perceived Social Green Preference: The Motivation Mechanism of Inducing Green Behaviour." *Current Psychology* 41: 1–17. https://doi.org/10.1007/s12144-021-02483-z.

Lin, Eden. 2019. "Why Subjectivists about Welfare Needn't Idealize." *Pacific Philosophical Quarterly* 100 (1): 2–23. https://doi.org/10.1111/papq.12232.

Lindgren, Ingrid, Anna Trulsson Schouenborg, Caroline Larsson, and Kjerstin Stigmar. 2023. "Perceptions of Everyday Life during Lenient COVID-19 Restrictions in Sweden—An Interview Study." *BMC Public Health* 23 (1): 1743. https://doi.org/10.1186/s12889-023-16599-3.

Lindsey, Rebecca. 2022. "Climate Change. Global Sea Level." *Climate Gov*, April 19, 2022. https://www.climate.gov/news-features/understanding-climate/climate-change-global-sea-level.

LisaLisa. 2019. "6 Reasons Why Second-Hand Clothes Are a Better Fashion Choice." *Night Helper*. https://nighthelper.com/6-reasons-second-hand-clothes-better-fashion-choice/.

Liu, Huijun, Shuzhuo Li, Qunying Xiao, and M. W. Feldman. 2014. "Social Support and Psychological Well-Being under Social Change in Urban and Rural China." *Social Indicators Research* 119 (2): 979–96. https://doi.org/10.1007/s11205-013-0534-1.

Lu, Yi, Jianting Zhao, Xueying Wu, and Siu M. Lo. 2021. "Escaping to Nature during a Pandemic: A Natural Experiment in Asian Cities during the COVID-19 Pandemic with Big Social Media Data." *Science of the Total Environment* 777: 146092. https://doi.org/10.1016/j.scitotenv.2021.146092.

Lucas, Richard E. 2007a. "Adaptation and the Set-Point Model of Subjective Well-Being: Does Happiness Change after Major Life Events?" *Current Directions in Psychological Science* 16 (2): 75–9. https://doi.org/10.1111/j.1467-8721.2007.00479.x.

Lucas, Richard E. 2007b. "Long-Term Disability Is Associated with Lasting Changes in Subjective Well-Being: Evidence from Two Nationally Representative Longitudinal Studies." *J Pers Soc Psychol* 92 (4): 717–30. https://doi.org/10.1037/0022-3514.92.4.717.

Lukianoff, Greg, and Jonathan Haidt. 2018. *The Coddling of the American Mind: How Good Intentions and Bad Ideas Are Setting Up a Generation for Failure*. New York: Penguin Books.

Lurquin, John H., and Akira Miyake. 2017. "Challenges to Ego-Depletion Research Go beyond the Replication Crisis: A Need for Tackling the Conceptual Crisis." *Frontiers in Psychology* 8: 568. doi: 10.3389/fpsyg.2017.00568.

Lyubomirsky, Sonja. 2001. "Why Are Some People Happier than Others? The Role of Cognitive and Motivational Processes in Well-Being." *Am Psychol* 56 (3): 239–49. http://dx.doi.org/10.1037/0003-066X.56.3.239.

Lyubomirsky, Sonja 2011. "Hedonic Adaptation to Positive and Negative Experiences." In *The Oxford Handbook of Stress, Health, and Coping*, edited by S. Folkman, 200–24. New York: Oxford University Press.

MacKinnon, James B. 2021. *The Day the World Stops Shopping. How Ending Consumerism Gives Us a Better Life and a Greener World*. London: Vintage Publishing.

Malm, Andreas. 2021. *How to Blow Up a Pipeline: Learning to Fight in a World on Fire*. London: Verso.

Martin, Adrienne. 2013. *How We Hope*. Princeton, NJ: Princeton University Press.

Maslow, Abraham. 1954. *Motivation and Personality*. New York: Harper.

Matheson, David. 2020. "The Worthwhileness of Meaningful Lives." *Philosophia* 48: 313–24. https://doi.org/10.1007/s11406-019-00082-8.

Mau, Martin, Anne-Maj Fabricius, and Søren H. Klausen. 2022. "Keys to Well-Being in Older Adults during the COVID-19 Pandemic: Personality, Coping and Meaning."

Int J Qual Stud Health Well-Being 17 (1): 2110669. https://doi.org/10.1080/17482631.2022.2110669.

Max-Neef, Manfred A. 1991. *Human-Scale Development: Conception, Application and Further Reflection*. London: Apex.

Maxwell, Keely, Susan Julius, Anne Grambsch, Ann Kosmal, Libby Larson, and Nancy Sonti. 2018. "Ch. 11: Built Environment, Urban Systems, and Cities." In *Impacts, Risks, and Adaptation in the United States: Fourth National Climate Assessment, Volume II*, U.S. Global Change Research Program, 438–78. Washington, DC. https://doi.org/10.7930/NCA4.2018.CH11.

McCaul, Kevin D., and John T. Kopp. 1982. "Effects of Goal Setting and Commitment on Increasing Metal Recycling." *Journal of Applied Psychology* 67 (3): 377–9. https://doi.org/10.1037/0021-9010.67.3.377.

McCrae, Robert R., and Juri Allik, eds. 2002. *The Five-Factor Model of Personality across Cultures*. Dordrecht: Kluwer Academic/Plenum Publishers. https://doi.org/10.1007/978-1-4615-0763-5.

McLennan, Jason F. 2014. "At What Point Does Size Cancel Out Sustainability?" *Green Home Gnome*, February 19, 2014. https://www.greenhomegnome.com/house-size-sustainability/.

McPherson, Tristram. 2021. "The Puzzle of Inefficacy." In *Philosophy and Climate Change*, edited by Mark Budolfson, Tristram McPherson, and David Plunkett (Oxford, 2021; online edn, Oxford Academic, April 22, 2021). Oxford. https://doi.org/10.1093/oso/9780198796282.003.0011, accessed 7 October 2025.

McQueen, Alison. 2021. "The Wages of Fear?: Toward Fearing Well about Climate Change." In *Philosophy and Climate Change*, edited by Mark Budolfson, Tristram McPherson, and David Plunkett, 152–77. Oxford University Press. https://doi.org/10.1093/oso/9780198796282.003.0008.

Merritt, Maria. 2000. "Virtue Ethics and Situationist Personality Psychology." *Ethical Theory and Moral Practice* 3 (4): 365–83.

Mertens, Stephanie, Mario Herberz, Ulf J. J. Hahnel, and Tobias Brosch. 2022. "The Effectiveness of Nudging: A Meta-Analysis of Choice Architecture Interventions across Behavioral Domains." *Proceedings of the National Academy of Sciences* 119: e2107346118. 10.1073/pnas.2107346118.

Messersmith, Emily E., and John E. Schulenberg. 2010. "Goal Attainment, Goal Striving, and Well-Being during the Transition to Adulthood: A Ten-Year U.S. National Longitudinal Study." *New Directions for Child and Adolescent Development* 2010 (130): 27–40. https://doi.org/10.1002/cd.279.

Metz, Thaddeus. 2012. "The Meaningful and the Worthwhile: Clarifying the Relationships." *The Philosophical Forum* 43: 435–48. https://doi.org/10.1111/j.1467-9191.2012.00436.x.

Metz, Thaddeus. 2013. *Meaning in Life: An Analytic Studio*. Oxford: Oxford University Press.

Meyer, David. 2017. "Congestion Pricing Was Unpopular in Stockholm, Until People Saw It in Action." Streetblog NYC, November 28, 2017. https://nyc.streetsblog.org/2017/11/28/congestion-pricing-was-unpopular-in-stockholm-until-people-saw-it-in-action/.

Mill, John Stuart. 1977. "*On Liberty*." In *The Collected Works of John Stuart Mill, Volume XVIII—Essays on Politics and Society Part I*. London: Routledge and Kegan Paul.

Millgram, Elijah. 2004. "IX—On Being Bored Out of Your Mind." *Proceedings of the Aristotelian Society* 104 (Hardback): 165–86. https://doi.org/10.1111/j.0066-7373.2004.00087.x.

Molouki, Sarah, Stephanie Y. Chen, Oleg Urminsky, and Daniel M. Bartels. 2020. "How Personal Theories of the Self Shape Beliefs about Personal Continuity and Transformative Experience." In *Becoming Someone New: Essays on Transformative Experience Choice and Change*, edited by Enoch Lambert and John Schwenkler, 52–72. Oxford: Oxford University Press.

Moncaster, Alice. 2021. "The Impact of the Built Environment on Climate Change—and of Climate Change on the Built Environment." *The Open University*, July 13, 2021. https://www.open.ac.uk/blogs/design/the-impact-of-the-built-environment-on-climate-change-and-of-climate-change-on-the-built-environment/.

Moore, Michael S. 2019. "The Rationality of Threshold Deontology." In *Moral Puzzles and Legal Perspectives*, edited by H. Hurd, 371–87. Cambridge: Cambridge University Press.

Morse, K. F., Philip A. Fine, and Kathryn J. Friedlander. 2021. "Creativity and Leisure during COVID-19: Examining the Relationship between Leisure Activities, Motivations, and Psychological Well-Being." *Frontiers in Psychology* 12: 609967. https://doi.org/10.3389/fpsyg.2021.609967.

Moscardo, Gianna, Alastair M. Morrison, Philip L. Pearce, Cheng-Te Lang, and Joseph T. O'Leary. 1996. "Understanding Vacation Destination Choice through Travel Motivation and Activities." *Journal of Vacation Marketing* 2 (2): 109–22. https://doi.org/10.1177/135676679600200202.

Moser, Stephanie, and Silke Kleinhückelkotten. 2018. "Good Intents, but Low Impacts: Diverging Importance of Motivational and Socioeconomic Determinants Explaining Pro-Environmental Behavior, Energy Use, and Carbon Footprint." *Environment and Behavior* 50 (6): 626–56. https://doi.org/10.1177/0013916517710685.

Mulgan, Tim. 2015. "Utilitarianism for a Broken World." *Utilitas* 27 (1): 92–114. https://doi.org/10.1017/S0953820814000338.

Nefsky, Julia. 2021. "Climate Change and Individual Obligations a Dilemma for the Expected Utility Approach, and the Need for an Imperfect View Julia Nefsky." In *Philosophy and Climate Change*, edited by Mark Budolfson, Tristram McPherson, and David Plunkett, 201–21. Oxford: Oxford University Press.

Negri, Marta, and Georg Bieker. 2025. "Life-Cycle Greenhouse Gas Emissions from Passenger Cars in the European Union." International Council on Clean Transportation.

Nelson, Eric S. 2020. *Daoism and Environmental Philosophy: Nourishing Life*. London: Routledge. https://doi.org/10.4324/9780429399145.

Nguyen, C. Thi. 2020. *Games: Agency as Art*. Oxford: Oxford University Press.

Nielsen, Camilla, Jaap Daalhuizen, and Philip Cash. 2017. "Behavioural Design Workbook—An Introduction to Behavioral Design and Its Application Copenhagen Commune." http://backend.orbit.dtu.dk/ws/portalfiles/portal/158819009/Behavioural_design_workbook_an_introduction_to_behavioral_design_and_its_application_in_Copenhagen_Commune.pdf.

Nielsen, Kristian Steensen. 2017. "From Prediction to Process: A Self-Regulation Account of Environmental Behavior Change." *Journal of Environmental Psychology* 51: 189–98.

Nielsen, Kristian Steensen, and Jan Michael Bauer. 2018. "The Merits of Goal Support as a Self- Control Strategy." *Social Psychological and Personality Science*. https://doi.org/10.1177/1948550618780729.

Nielsen, Kristian Steensen, and Wilhelm Hofmann. 2021. "Motivating Sustainability through Morality: A Daily Diary Study on the Link between Moral Self-Control and Clothing Consumption." *Journal of Environmental Psychology* 73: 101551.

Nielsen, Lasse. 2019. "Sufficiency and Satiable Values." *Journal of Applied Philosophy* 36 (5): 800–16. https://doi.org/https://doi.org/10.1111/japp.12364.

Nozick, Robert. 1974. *Anarchy, State, and Utopia*. New York: Basic Books.

Nozick, Robert. 1989. *The Examined Life*. New York: Simon & Schuster.

Nungesser, Frithjof, and Martin Winter. 2021. "Meat and Social Change: Sociological Perspectives on the Consumption and Production of Animals." *OZS Osterr Z Soziol* 46 (2): 109–24. https://doi.org/10.1007/s11614-021-00453-0.

Nussbaum, Martha C. 2009. "Creating Capabilities: The Human Development Approach and Its Implementation." *Hypatia* 24 (3): 211–15. https://doi.org/https://doi.org/10.1111/j.1527-2001.2009.01053.x.

O'Neill, Daniel W. 2020. "Beyond Green Growth." *Nature Sustainability* 3 (4): 260–1. https://doi.org/10.1038/s41893-020-0499-4.

O'Neill, John. 2017. "How Not to Argue against Growth: Happiness, Austerity and Inequality." In *The Good Life beyond Growth. New Perspectives*, edited by Hartmut Rosa and Christoph Henning, 141–52. London: Routledge. https://doi.org/10.4324/9781315542126.

OECD. 2023. "Green Growth and Sustainable Development." https://www.oecd.org/greengrowth/.

Orden-Mejía, Miguel, Mauricio Carvache-Franco, Assumpció Huertas, Wilmer Carvache-Franco, Nathalie Landeta-Bejarano, and Orly Carvache-Franco. 2022. "Post-COVID-19 Tourists' Preferences, Attitudes and Travel Expectations: A Study in Guayaquil, Ecuador." *Int J Environ Res Public Health* 19 (8): 4822. https://doi.org/10.3390/ijerph19084822.

Osler, Lucy 2021. "Taking Empathy Online." *Inquiry* 67 (1): 302–29. https://doi.org/10.1080/0020174X.2021.1899045.

Panarese, Paola, and Vittoria Azzarita 2021. "The Impact of the COVID-19 Pandemic on Lifestyle: How Young People Have Adapted Their Leisure and Routine during Lockdown in Italy." YOUNG 2021. 10.1177/11033088211031389.

Pandey, Sujitam Annemarie Olsen, Federico J. A. Perez-Cueto, and Marianne Thomsen. 2023. "Nudging toward Sustainable Food Consumption at University Canteens: A Systematic Review and Meta-Analysis." *Journal of Nutrition Education and Behavior* 55 (12): 894–904.

Parfit, Derek. 1982."Future Generations: Further Problems." *Philosophy & Public Affairs* 11 (2): 113–72.

Parfit, Derek. 1984. *Reasons and Persons*. Oxford University Press.

Parfit, Derek. 1986. *Reasons and Persons*. Oxford: Oxford University Press. https://doi.org/10.1093/019824908x.001.0001.

Pasqualetti, Martin J. 2011. "Opposing Wind Energy Landscapes: A Search for Common Cause." *Annals of the Association of American Geographers* 101 (4): 907–17. http://www.jstor.org/stable/27980237.

Paul, Laurie A. 2014. *Transformative Experience*. Oxford: Oxford University Press. https://doi.org/10.1093/acprof:oso/9780198717959.001.0001.

Pearson, James S. 2023. "Realism in the Ethics of Immigration." *Philosophy and Social Criticism* 49 (8): 950–74.

Peters, Terri, and Stephen Verderber. 2021. "Biophilic Design Strategies in Long-Term Residential Care Environments for Persons with Dementia." *Journal of Aging and Environment* 36 (3): 227–55. doi: 10.1080/26892618.2021.1918815.

Pettit, Philip. 2004. "Hope and Its Place in Mind." *The Annals of the American Academy of Political and Social Science* 592 (1): 152–65.

Pinker, Stephen. 2002. *The Blank Slate: The Modern Denial of Human Nature*. London: Viking.

Plato. 1992. *Republic*. 2 nd Edition. Translated by M. A. Grube, Rev. by C. D. C. Reeve. Indianapolis, IN: Hackett.

Pomponi, Francesco, Ruth Saint, Jay H. Arehart, Niaz Gharavi, and Bernadino D'Amico. 2021. "Decoupling Density from Tallness in Analysing the Life Cycle Greenhouse Gas Emissions of Cities." *NPJ Urban Sustainability* 1 (1): 33. https://doi.org/10.1038/s42949-021-00034-w.

Popper, Karl Raimond. (1957). *The Poverty of Historicism* (Vol. 9). London: Routledge and Kegan Paul.

Przybylski, Andrew K., C. Scott Rigby, and Richard M. Ryan. 2010. "A Motivational Model of Video Game Engagement." *Review of General Psychology* 14 (2): 154–66. https://doi.org/10.1037/a0019440.

Rabbiosi, Chiara. 2014. "The Condition of 'Cosmo-housewives': Leisure Shopping, the Mainstream and Its Ambiguities." *Gender, Place & Culture* 21 (2): 211–27. https://doi.org/10.1080/0966369X.2013.791250.

Railton, Peter. 2021. "A Convenient Truth?: Climate Change and Quality of Life." In *Philosophy and Climate Change*, edited by Mark Budolfson, Tristram McPherson, and David Plunkett, 9–41. Oxford: Oxford University Press. https://doi.org/10.1093/oso/9780198796282.003.0002.

Rauschmayer, Felix, and Ines Omann. 2015. "Well-Being in Sustainability Transitions—Making Use of Needs." In *Sustainable Consumption and the Good Life—Interdisciplinary Perspectives*, edited by Karen Syse and Martin Mueller, 111–25. London: Routledge. https://doi.org/10.4324/9781315795522.

Rawls, John 1971. *A Theory of Justice*. Cambridge, MA: Harvard University Press.

Raz, Joseph. 1986. *The Morality of Freedom*. Oxford: Clarendon Press.

Redclift, Michael 2005. "Sustainable Development (1987–2005): An Oxymoron Comes of Age." *Sustainable Development* 13: 212–27.

Redmond, Brian F. 2010. *Self-Efficacy Theory: Do I Think That I Can Succeed in My Work? Work Attitudes and Motivation*. University Park, PA: Pennsylvania State University Press.

Rhoads, Shawn A., and Abigail Marsh. 2023. "Doing Good and Feeling Good: Relationships between Altruism and Well-Being for Altruists, Beneficiaries, and Observers." *World Happiness Report 2023* (11th edition, Chapter 4). New York: Sustainable Development Solutions Network.

Rial, Rafael Cardoso. 2024. "Biofuels versus Climate Change: Exploring Potentials and Challenges in the Energy Transition." *Renewable and Sustainable Energy Reviews* 196. https://www.sciencedirect.com/science/article/pii/S1364032124000923.

Ridder, Denise T. D., Gerty Lensvelt-Mulders, Catrin Finkenauer, F. Marijn Stok, and Roy F. Baumeister. 2012. "Taking Stock of Self-Control: A Meta-Analysis of How Trait Self-Control Relates to a Wide Range of Behaviors." *Personality and Social Psychology Review* 16 (1): 76–99. https://doi.org/10.1177/1088868311418749.

Rittel, Horst, and Melvin Webber. 1973. "Dilemmas in a General Theory of Planning." *Policy Sciences* 4 (2): 155–69.

Robeyns, Ingrid and Morten F. Byskov. 2020. "The Capability Approach." In *The Stanford Encyclopedia of Philosophy* (Winter 2020 Edition), edited by Edward N. Zalta. https://plato.stanford.edu/archives/win2020/entries/capability-approach/.

Robinson, Kim S. 2020. *The Ministry for the Future*. New York: Orbit.

Rosa, Hartmut. 2017. "Available, Accessible, Attainable. The Mindset of Growth and the Resonance Conception of the Good Life." In *The Good Life beyond Growth: New Perspectives*, edited by Hartmut Rosa and Chrstoph Henning, 39–55. London: Routledge.

Russell, Bertrand. 1930. *The Conquest of Happiness*. London: Liveright.

Rutledge, Garrett. 2020. "Life in 2050. A Look into Sustainable Cities of the Future." *New Jersey Digest*, October 22, 2020. https://thedigestonline.com/community-human-interest/sustainable-cities-of-the-future/.

Ryan, Richard M., and Edward L. Deci. 2017. *Self-Determination Theory: Basic Psychological Needs in Motivation, Development, and Wellness.* New York: Guilford Publications. https://books.google.dk/books?id=GF0ODQAAQBAJ.

Ryle, Martin, and Kate Soper. 2002. *To Relish the Sublime? Culture and Self-Realization in Postmodern Times.* London: Verso.

Ryu, Airi. 2023. "Ranked: Numbers of Agricultural Workers by Country." *Visual Capitalist*, August 3, 2023. https://www.visualcapitalist.com/cp/agricultural-workers-by-country/.

Sahlins, Marshall. 1972. *Stone Age Economics.* New York: Aldine-Atherton.

Savickas, Mark L. 2013. "Career Construction Theory and Practice." In *Career Development and Counseling: Putting Theory and Research to Work*, edited by S. D. Brown and R. W. Lent, 147–83. Hoboken, NJ: John Wiley & Sons Inc.

Scanlon, Thomas. 1975. "Thomson on Privacy." *Philosophy and Public Affairs* 4 (4): 315–22.

Schelling, Thomas C. 1971. "Dynamic Models of Segregation." *The Journal of Mathematical Sociology* 1 (2): 143–86. doi: 10.1080/0022250X.1971.9989794.

Schiller, Friedrich. 1982. *On the Aesthetic Education of Man in a Series of Letters.* Translated by Elizabeth M. Wilkinson and L. A. Willoughby. Oxford: Oxford University Press.

Schiller, F. 2005. "On Grace and Dignity" J. V. Curran, translated by In *Schiller's "On Grace and Dignity" in Its Cultural Context: Essays and a New Translation*, edited by J. V. Curran and C. Fricker, 123–70. Rochester, NY: Camden House.

Schneider, François, Joan Martinez-Alier, and Giorgos Kallis. 2011. "Sustainable Degrowth." *Journal of Industrial Ecology* 15 (5): 654–6. https://doi.org/10.1111/j.1530-9290.2011.00388.x.

Schramme, Thomas. 2013. "Disability (Not) as a Harmful Condition: The Received View Challenged." In *Disability and the Good Human Life*, edited by Jerome E. Bickenbach, Franziska Felder, and Barbara Schmitz, 72–92. Cambridge: Cambridge University Press. https://doi.org/DOI:10.1017/CBO9781139225632.004.

Schütz, Astrid, and Roy F. Baumeister. 2017. "Positive Illusions and the Happy Mind." In *The Happy Mind: Cognitive Contributions to Well-Being*, edited by Michael D. Robinson and Michael Eid, 177–93. Cham: Springer.

Schwartz, Barry. 2004. *The Paradox of Choice: Why More Is Less.* New York: Harper Collins.

Schweizer, S., R. P. Lawson, and S. J. Blakemore. 2023. "Uncertainty as a Driver of the Youth Mental Health Crisis." Current Opinion in Psychology 53: 101657. https://doi.org/10.1016/j.copsyc.2023.101657.

Schwitzgebel, Eric. 2022. "The Nature of Belief from a Philosophical Perspective, with Theoretical and Methodological Implications for Psychology and Cognitive Science." *Frontiers in Psychology* 13: 947664. https://doi.org/10.3389/fpsyg.2022.947664. https://www.frontiersin.org/articles/10.3389/fpsyg.2022.947664.

Scott, James C. (1998). *Seeing Like a State: How Certain Schemes to Improve the Human Condition Have Failed*. New Haven, CT: Yale University Press.

Scott, James C. 2017. *Against the Grain: A Deep History of the Earliest States*. New Haven, CT: Yale University Press.

Searle, John R. 1992. *The Rediscovery of the Mind*. Cambridge, MA: MIT Press.

Sen, Amartya. 2000. *Development as Freedom*. 1st Anchor Books Edition. New York: Anchor Books. https://search.library.wisc.edu/catalog/999977297202121.

Sharot, Tali, and Cass R. Sunstein. 2024. *Look Again: The Power of Noticing What Was Always There*. Dallas: Brown Books.

Sheldon, Kennon M. 2008. "Assessing the Sustainability of Goal-Based Changes in Adjustment over a Four-Year Period." *Journal of Research in Personality* 42 (1): 223–9. https://doi.org/10.1016/j.jrp.2007.03.002.

Sheldon, Kennon M., and Christopher P. Niemiec. 2006. "It's Not Just the Amount That Counts: Balanced Need Satisfaction Also Affects Well-Being." *Journal of Personality and Social Psychology* 91 (2): 331–41. https://doi.org/10.1037/0022-3514.91.2.331.

Sheldon, Kennon M., and Richard E. Lucas. 2014. "Is It Possible to Become A Permanently Happier Person? An Overview of the Issues and the Book." In *Stability of Happiness*, edited by Kennon M. Sheldon and Richard E. Lucas, 1–7. Cambridge, MA: Academic Books. https://doi.org/10.1016/B978-0-12-411478-4.00001-1.

Sheldon, Kennon M., and Sonja Lyubomirsky. 2012. "The Challenge of Staying Happier: Testing the Hedonic Adaptation Model." *Personality and Social Psychology Bulletin* 38 (5): 670–80. https://doi.org/10.1177/0146167212436400.

Sheldon, Kennon M., Julia Boehm, and Sonja Lyubomirsky. 2012. "Variety Is the Spice of Happiness: The Hedonic Adaptation Prevention Model." In *Oxford Handbook of Happiness* (online edition), edited by Ilona Boniwell, Susan A. David, and Amanda C. Ayers, 901–14. Oxford: Oxford University Press. https://doi.org/10.1093/oxfordhb/9780199557257.013.0067.

Sherman, Rachel. 2017. "What the Rich Won't Tell You." *International New York Times*, September 8, 2017, NA. *Gale Academic OneFile* (accessed April 8, 2024). https://link.gale.com/apps/doc/A504075785/AONE?u=anon~268bbb46&sid=sitemap&xid=0c73b20c.

Shriver, Adam J. 2014. "The Asymmetrical Contributions of Pleasure and Pain to Animal Welfare." *Cambridge Quarterly of Healthcare Ethics* 23 (2): 152–62. https://doi.org/10.1017/S0963180113000686.

Simmel, Georg. 1997. "Sociology of the Meal." In *Simmel on Culture: Selected Writings*, edited by David Patrick Frisby and Mike Featherstone, 130–5. London, Thousand Oaks, CA, and New Delhi: SAGE.

Singer, Peter. 2010. "A Darwinian Left: Politics, Evolution, and Cooperation." In *Philosophy after Darwin*, edited by Michael Ruse, 343–9. Princeton, NJ: Princeton University Press. https://doi.org/doi:10.1515/9781400831296-039.

Sinnott-Armstrong, Walter, 2005. "It's Not My Fault: Global Warming and Individual Moral Obligations." In *Perspectives on Climate Change*, edited by Walter Sinnott-Armstrong and Richard B. Howarth, 221–53. Amsterdam: Elsevier.

Sjöblom, Pia, Lili-Ann Wolff, and Jessica Sundman. 2024. "Climate Change as a Socio-Scientific Issue in Upper Secondary Education: Addressing Wicked Problems through Cross-Curricular Approaches." In *Developing a Didactic Framework across and beyond School Subjects*, edited by Søren H. Klausen and Nina Mård, 182–96. London: Routledge.

Solomon, Robert C. 1993. *The Passions: Emotions and the Meaning of Life*. Indianapolis, PA: Hackett Publishing.

Song, Yujia. 2017. "The Moral Virtue of Open-Mindedness." *Canadian Journal of Philosophy* 48 (1): 65–84.

Soper, Kate. 2020. *Post-Growth Living. For an Alternative Hedonism*. New York: Verso.

Sousa, Ronald de. 1987. *The Rationality of Emotion*. Cambridge, MA: MIT Press.

Spiekermann, Kai. 2014. "Small Impacts and Imperceptible Effects." *Midwest Studies In Philosophy* 38: 75–90.

Stace, Walter T. 1944. "Interestingness." *Philosophy* 19 (74): 233–41. https://doi.org/10.1017/S003181910002194X.

Stauner, Nick. 2013. *Personal Goal Attainment, Psychological Well-Being Change, and Meaning in Life*. UC Riverside Electronic Theses and Dissertations. http://escholarship.org/uc/item/3t34c68w.

Strawson, Galen 2004. "Against Narrativity." *Ratio* 17: 428–52.

Strohminger, Nina, Joshua Knobe, and George Newman. 2017. "The True Self: A Psychological Concept Distinct from the Self." *Perspectives on Psychological Science: A Journal of the Association for Psychological Science* 12 (4): 551560. https://doi.org10.1177/1745691616689495.

Strömbeck, Camilla, Emma Lindkvist, and Daniel Västfjäl. 2023. "Individual Differences in Environmental Wellbeing and Pro-environmental Behaviors Explained by Self-Control." *Frontiers in Psychology* 14: 1088682.

Sullivan, Meghan. 2018. *Time Biases: A Theory of Rational Planning and Personal Persistence*. Oxford: Oxford University Press.

Sunstein, Cass R. 2019. *How Change Happens*. Cambridge, MA: MIT Press.

Swanton, Christine. 2021 *Target Centred Virtue Ethics*. Oxford: Oxford University Press.

Taufik, Danny, Jan W. Bolderdijk, and Linda Steg. 2015. "Acting Green Elicits a Literal Warm Glow." *Nature Climate Change* 5: 37–40.

Thaler, Richard H., and Cass R. Sunstein. 2008. *Nudge: Improving Decisions about Health, Wealth, and Happiness*. New Haven, CT: Yale University Press.

Tiberius, Valerie. 2015. "Prudential Value." In *The Oxford Handbook of Value Theory*, edited by Iwao Hirose and Jonas Olson, 158–74. Oxford: Oxford University Press. https://doi.org/10.1093/oxfordhb/9780199959303.013.0010.

Tiberius, Valerie. 2018. *Well-Being as Value Fulfillment: How We Can Help Each Other to Live Well*. Oxford: Oxford University Press. https://doi.org/10.1093/oso/9780198809494.001.0001.

Tobia, Kevin P. 2015. "Personal Identity and the Phineas Gage Effect." *Analysis* 75 (3): 396–405. https://doi.org/10.1093/analys/anv041.

Tolstoy, Leo. 2016. *War and Peace*. Translated by Anthony Briggs. London: Penguin Books.

UN (2023). *The Sustainable Development Goals Report 2023*. https://unstats.un.org/sdgs/report/2023/The-Sustainable-Development-Goals-Report-2023.pdf.

UN. 2023a. "The Sustainable Development Agenda." https://www.un.org/sustainabledevelopment/development-agenda/.

UN. 2023b. "Why Sustainable Lifestyles Matter." https://www.unep.org/explore-topics/resource-efficiency/what-we-do/sustainable-lifestyles/why-sustainable-lifestyles.

United States Environmental Protection Agency. 2023. "Heat Island Effect." https://www.epa.gov/heatislands.

University of Minnesota Libraries (authors anonymous). 2016. *Sociology: Understanding and Changing the Social World*. University of Minnesota Libraries Publishing. https://doi.org/10.24926/8668.2401.

Urminsky, Oleg. 2017. "The Role of Psychological Connectedness to the Future Self in Decisions over Time." *Current Directions in Psychological Science* 26 (1): 34–9. https://doi.org/10.1177/0963721416668810.

van Dellen, Michelle R., James Y. Shah, N. Pontus Leander, Julie E. Delose, and Jerica X. Bornstein. 2015. "In Good Company. Managing Interpersonal Resources That Support Self- Regulation." *Personality and Social Psychology Bulletin* 41 (6): 869–82.

van der Deijl, Willem J. A. 2020. "A Challenge for Capability Measures of Wellbeing." *Social Theory and Practice* 46 (3): 605–31. https://doi.org/10.5840/soctheorpract202071799.

van der Linden, Sandor. 2018. "Warm Glow Is Associated with Low- but Not High-cost Sustainable Behaviour." *Nature Sustainability* 1: 28–30.

Van Rookhuijzen, Marije, Emily De Vet, and Marieke Adriaanse. 2021. "The Effects of Nudges: One-Shot Only? Exploring the Temporal Spillover Effects of a Default Nudge." *Frontiers in Psychology* 12: 683262. doi: 10.3389/fpsyg.2021.683262.

van Valkengoed, Anne M., Wokje Abrahamse, and Linda Steg. 2022. "To Select Effective Interventions for Pro-environmental Behaviour Change, We Need to Consider Determinants of Behaviour." *Nature Human Behavior* 6: 1482–92. https://doi.org/10.1038/s41562-022-01473-w.

van Wijngaarden, E., C. Leget, and A. Goossensen. 2015. "Ready to Give Up on Life: The Lived Experience of Elderly People Who Feel Life Is Completed and No Longer Worth Living." *Social Science & Medicine* 138: 257–64.

VanderWeele, T. J. 2017. "On the Promotion of Human Flourishing." *Proceedings of the National Academy of Sciences of the United States of America* 114 (31): 8148–56.

Vacek, Z., S. Vacek, J. Cukor. 2023 "European Forests under Global Climate Change: Review of Tree Growth Processes, Crises and Management Strategies." *J Environ Manage* April 15, 332: 117353. doi: 10.1016/j.jenvman.2023.117353. Epub January 28, 2023. PMID: 36716544.

Varga, S. 2011. *Authenticity as an Ethical Ideal*. London: Routledge.

Vásquez-Levy, D. 2002. "Bildung-Centred Didaktik: A Framework for Examining the Educational Potential of Subject Matter." *Journal of Curriculum Studies* 34 (1): 117–28.

Velleman, J. David. 1991. "**Well-Being and Time**." *Pacific Philosophical Quarterly* 72 (1): 48–77.

Vendrell Ferran, Ingrid. 2021. "How to Understand Feelings of Vitality: An Approach to Their Nature, Varieties, and Functions." In *Phenomenology of Bioethics: Technoethics and Lived-Experience*, edited by Susi Ferrarello, 115–30. The International Library of Bioethics, vol 84. Cham: Springer.

Vousdoukas, Michalis I., Lorenzo Mentaschi, Evangelos Voukouvalas, Alessandra Bianchi, Fransesco Dottori, and Luc Feyen. 2018. "Climatic and Socioeconomic Controls of Future Coastal Flood Risk in Europe." *Nature Climate Change* 8 (9): 776–80. https://doi.org/10.1038/s41558-018-0260-4.

Vyas, Lina. 2022. "'New Normal' at Work in a Post-COVID World: Work–Life Balance and Labor Markets." *Policy and Society* 41 (1): 155–67. https://doi.org/10.1093/polsoc/puab011.

Wallace-Wells, David. 2019. *The Uninhabitable Earth: A Story of the Future*. London: Penguin.

Wang, Qing, and José A. Tapia Granados. 2019. "Economic Growth and Mental Health in 21st Century China." *Social Science & Medicine* 220: 387–95. https://doi.org/https://doi.org/10.1016/j.socscimed.2018.11.031.

Waterman, Allan S., Seth J. Schwartz, and Regina Conti. 2008. "The Implications of Two Conceptions of Happiness (Hedonic Enjoyment and Eudaimonia) for the Understanding of Intrinsic Motivation." *Journal of Happiness Studies* 9 (1): 41–79. https://doi.org/10.1007/s10902-006-9020-7.

Watts, Duncan J. 2011. *Everything Is Obvious, Once You Know the Answer: How Common Sense Fails Us*. New York: Crown Business/Random House.

Weimann, Joachim, Andreas Knabe, and Ronnie Schöb. 2015. *Measuring Happiness: The Economics of Well-Being*. Cambridge, MA: The MIT Press. https://doi.org/10.7551/mitpress/10036.001.0001.

Weingarten, Oliver. 2015. "'Self-Cultivation' (xiu shen 修身) in the Early Edited Literature." *Oriens Extremus* 54: 163–206.

Welsch, Heinz, 2020. "Happiness and Green Lifestyle." In *Handbook on Wellbeing, Happiness and the Environment*, edited by David Maddison, Katrin Rehdanz, and Heinz Welsch, chapter 19, 349–66. Cheltenham: Edward Elgar Publishing.

Welsch, Heinz, Martin Binder, and Ann-Kathrin Blankenberg. 2021. "Green Behavior, Green Self-Image, and Subjective Well-Being: Separating Affective and Cognitive Relationships." *Ecological Economics* 179. https://doi.org/10.1016/j.ecolecon.2020.106854.

Wendel, S. 2020. *Designing for Behavior Change*. 2nd Edition. San Francisco, CA: O'Reilly Media, Inc.

White, Sarah C. 2017. "Relational Wellbeing: Re-centring the Politics of Happiness, Policy and the Self." *Policy & Politics* 45 (2): 121–36. https://doi.org/10.1332/030557317X14866576265970.

Willett, Walter, Johan Rockström, Brent Loken, Marco Springmann, Tim Lang, Sonja Vermeulen, Tara Garnett, David Tilman, Fabrice DeClerck, Amanda Wood Malin Jonell, Michael Clark, Line J. Gordon, Jessica Fanzo, Prof Corinna Hawkes, Rami Zurayk, Juan A. Rivera, Wim De Vries, Lindiwe Majele Sibanda, Ashkan Afshin, Abhishek Chaudhary, Mario Herrero, Rina Agustina, Francesco Branca, Anna Lartey, Shenggen Fan, Beatrice Crona, Elizabeth Fox, Victoria Bignet, Max Troell, Therese Lindahl, Sudhvir Singh, Sarah E. Cornell, K. Srinath Reddy, Sunita Narain, Sania Nishtar, and Christopher J. L. Murray. 2019. "Food in the Anthropocene: the EAT–Lancet Commission on Healthy Diets from Sustainable Food Systems." *The Lancet* 393 (10170): 447–92. https://doi.org/10.1016/S0140-6736(18)31788-4.

Williams, Bernard. 1973. "The Makropulos Case: Reflections on the Tedium of Immortality." In *Problems of the Self: Philosophical Papers 1956-1972*, 82–100. Cambridge: Cambridge University Press. https://doi.org/DOI:10.1017/CBO9780511621253.008.

Wilson, Timothy D., and Daniel T. Gilbert. 2008. "Explaining Away: A Model of Affective Adaptation." *Perspectives on Psychological Science: A Journal of the Association for Psychological Science* 3 (5): 370–86.

Wilterdink, Nico, William Form. 2025. "Social Structure." *Encyclopedia Britannica*, September 6, 2025. https://www.britannica.com/topic/social-structure. Accessed 6 October 2025.

Wittgenstein, Ludwig. 1922. *Tractatus Logico-Philosophicus*. Translated by Charles K. Ogden. London: Kegan Paul.

Wolf, Susan. 2012. *Meaning in Life and Why It Matters*. Princeton: Princeton University Press.

Wood, Wendy. 2019. *Good Habits, Bad Habits: The Science of Making Positive Changes that Stick*. New York: Farrar, Straus and Giroux.

World Economic Forum. 2021. "Are Electric Cars All That Green? Yes, This Chart Shows You How." https://www.weforum.org/agenda/2021/08/electric-cars-emissions-climate-change-green-environment/.

World Economic Forum. 2022. "How many jobs could the green energy transition create?". https://www.weforum.org/stories/2022/03/the-clean-energy-employment-shift-by-2030/

Wu, Tsaiyi. 2020. "A Dream of a Stone: The Ethics of De-anthropocentrism." *Open Philosophy* 3 (1): 413–28. https://doi.org/10.1515/opphil-2020-0104.

Wündisch, Joachim. 2014. "Green Votes Not Green Virtues: Effective Utilitarian Responses to Climate Change." *Utilitas* 26 (2): 192–205. https://doi.org/10.1017/S0953820813000307.

Zaleski, Zbigniew, Malgorzata Sobol-Kwapinska, Aneta Przepiorka, and Michal Meisner. 2019. "Development and Validation of the Dark Future Scale." *Time Soc.* 28: 107–23.

Index of Names

Alexandrova, A. 97
Appiah, A. 129, 146
Aristotle 25, 53, 141, 151

Brülde, B. 10, 36, 49

Calhoun, C. 28, 100, 102, 105, 110, 122, 173
Callard, A. 104, 144, 175

Doris, D. 139, 146

Elster, J. 61, 83

Goethe, J. W. v. 27, 140, 142
Graeber, D. 24, 26
Greif, M. 26

Haybron, D. 52, 56, 96, 119, 178
Humboldt, W. v. 50, 112, 115, 118, 140, 142, 164

Jamieson, D. 5, 8, 11, 163

Kant, I. 10, 141, 157
Kierkegaard, S. 89, 108, 115, 117, 123, 140, 179

Max-Neef, M. 67, 71
Mill, J. S. 163
Mulgan, T. 59, 62, 167

Nielsen, L. 68, 72
Nozick, R. 28, 57, 100
Nussbaum, M. 10

O'Neill, D. W. 12, 14, 57, 60, 79

Paul, L. 11, 28, 55, 100, 103
Plato 25, 136

Railton, P. 9–10
Rawls, J. 86
Raz, J. 68, 72
Rosa, H. 10, 58

Schiller, F. 140, 145, 149, 157, 179
Sen, A. 10, 60, 79, 95
Soper, K. 10, 89, 115, 151
Sunstein, C. 104, 114, 117, 154, 160

Tiberius, V. 51, 91
Tolstoy, L. 120

Wood, W. 9, 13, 131, 168

Index of Subjects

adaptive preferences 58, 83
aesthetics 177, 197
anthropocentrism 22, 143, 155
aspiration 175, 193
autonomy 75, 141, 151, 159, 162

Bildung 137, 175, 192, 199

change, impact on wellbeing 99
change, measure of 29
change and stability, value of 27, 99, 165, 179, 200
climate intervention ethics 5, 159
commitment 101, 177, 188, 199
Confucianism 151

environmental humanities 5, 121, 155, 178

feasibility 169, 171, 181, 189
flourishing 8, 53, 70, 75, 112, 145
freedom 67, 96, 114, 160, 163

growth 12, 57, 64, 94, 165, 200
growth, green 11–12, 30, 43

harm principle 161
hedonic adaptation 58, 79, 110, 189
hedonism 8, 10, 52, 75
hope 171, 189

individualism 63, 143

liberalism 63, 160, 162

life satisfaction 53, 79

meaning in life 105

nature 5, 155
needs 10, 51, 62, 85, 94, 166, 195, 200
nudging 104, 137, 160

posthumanism 155
preference satisfaction 52, 75, 93

repetition 132, 135, 154, 188, 199

satiable values and needs 63, 68, 78
self-control 131, 142, 148, 155
self-cultivation 5, 21, 105, 129, 139, 150, 154
self-determination theory 9, 74, 130
social contagion 48, 152, 185

tipping points 153, 171, 188

transformative experience 101, 192

variation 9, 48, 62, 86, 89, 112, 146, 170, 199
virtues 13, 108, 139, 145

wellbeing, objective 9, 54, 65, 82, 108
wellbeing, process-oriented theory 95, 110
wellbeing, subjective 9, 10, 21, 54, 57, 70, 74, 79, 84, 94, 130, 167, 179, 195

About the Author

Søren Harnow Klausen is professor of philosophy at the University of Southern Denmark. His wide interests span analytical and continental philosophy (especially phenomenology and existential philosophy), philosophy of mind, epistemology (especially social epistemology), ethics and interdisciplinary work on transition processes, new technology, wellbeing, health, and aging. He was visiting scholar at Peking University 2023–4 and at Fudan University (Shanghai) 2024 and 2025. His publications include "Group Knowledge: A Real-World Approach" (*Synthese,* 2015); "Ethics, Knowledge, and a Procedural Approach to Wellbeing" (*Inquiry,* 2018); *Kierkegaard: Educating for Authenticity* (Springer, 2018); Perspectives *on Happiness* (Brill, 2019); and "The Phenomenology of Aging" (*Phenomenology and the Cognitive Sciences*, 2024). He is currently head of the research project *Wellbeing and Virtual Worlds* funded by Independent Research Fund Denmark.

www.ingramcontent.com/pod-product-compliance
Ingram Content Group UK Ltd.
Pitfield, Milton Keynes, MK11 3LW, UK
UKHW020212290926
15470UKWH00001B/145

9 781666 914412